Electroanalytical Stripping Methods

CHEMICAL ANALYSIS

A SERIES OF MONOGRAPHS ON ANALYTICAL CHEMISTRY AND ITS APPLICATIONS

Editor
J. D. WINEFORDNER

VOLUME 126

A WILEY-INTERSCIENCE PUBLICATION

JOHN WILEY & SONS

New York / Chichester / Brisbane / Toronto / Singapore

Electroanalytical Stripping Methods

Kh. BRAININA

Ural Institute of National Economy
Ekaterinburg, Russia

E. NEYMAN

Central Special Inspection of Russian Environmental
Protection Committee
Moscow, Russia

A WILEY-INTERSCIENCE PUBLICATION

JOHN WILEY & SONS

New York / **Chichester** / **Brisbane** / **Toronto** / **Singapore**

Library of Congress Cataloging in Publication Data:

Brainina, Kh. (Khjena), 1930-
    Electroanalytical stripping methods / Kh. Brainina, E. Neyman.
      p.   cm.—(Chemical analysis ; v. 126)
    "A Wiley-Interscience publication."
    Includes bibliographical references.
    ISBN 0-471-59506-3
    1. Electrochemical analysis. I. Neyman, E. II. Title.
III. Series.
QD115.B655   1993
543′.0874—dc20   93-20423

To the bright memory of the late academician,
A. N. Frumkin

PREFACE

Analytical chemistry as a whole and electroanalytical methods in particular are connected with many fields of natural science and technology [1, 2]. This makes it necessary to consider two aspects of these methods: their development as a technique in the context of progress in cognate fields and their application in various fields of science and technology. Because of the wide range of these topics this book is limited mainly to stripping electroanalytical methods as a source of information on the concentration and some properties of substances; the fundamental problems of electrochemical theory and methodology have been adequately dealt with in books [3–5], and those of mathematical methods and instrumentation in papers [6–9].

Analyzing journals and the citation index of papers on analytical chemistry, Orient and co-workers [10, 11] have shown that more papers are concerned with electrochemical methods than any other topic, owing to the rapid development of the theory [12–16], deeper understanding of the mechanism and kinetics of electrode processes, development of instrumentation [4, 6, 12, 13], evolution of modified electrodes [12, 17], and, as a consequence, substantial broadening of the fields of application. This is because electroanalytical methods furnish, in particular, information inaccessible by other methods, for example, in medicine, biochemistry and molecular biology, monitoring of the environment [18–20], and investigation of solids [14, 21, 22]. Potential of the application of polarography and voltammetry as recommended methods were considered by Bersiers [8–9].

Electrochemical detectors [23] are important in various rapidly developing hybrid analytical methods [1] such as liquid chromatography and flow-injection analysis. Comparison of the detection limits is often in favor of voltammetry, and especially of stripping voltammetry. Recent advantages in stripping analysis were considered by Wang [24, 25].

The extraordinary sensitivity of stripping techniques, down to 10^{-10} M and less, shown for organic compounds and some toxic metal ions is due to trace amounts of material concentrated on a microelectrode by electrochemical deposition before the process is reversed and the material is stripped off the electrode. This idea expressed by Barker in 1952 [26] proved to be very fruitful.

Recent developments in ultratrace stripping analysis utilize an adsorptive approach. Stripping voltammetry at mercury or solid electrodes provides for very sensitive determination with detection limits in the $10^{-8} - 2.5 \times 10^{-11}$ M range for surface-active compounds which cannot be accumulated electrochemically [27].

The increased use of on-line stripping analysis is expected in the near future. Apart from the obvious advantages of on-line monitoring and possible automation (as desirable in many environmental, clinical, and industrial situations), such adaptations offer several advantages over stripping measurements in batch systems.

Analytical chemists are increasingly being called on to determine not only amounts of elements but also their chemical form, so that polarography and voltammetry find an ever-increasing application in, for example, environmental studies.

Electrochemical methods have some distinct advantages for speciation measurements: (i) Electrochemical methods can be highly sensitive, a feature which is essential when analyzing natural waters, where the total concentration of a heavy metal is often in the $\mu g \, L^{-1}$ or $ng \, L^{-1}$ range. (ii) Operating conditions can be adjusted so that only metals that form complexes whose rate of dissociation is within a desired range are included in the electroactive fraction [27].

In recent years several books on polarography [5, 16] and voltammetry [4, 14, 15, 28, 29] have been published. *Analytical Chemistry* publishes biennial reviews of dynamic electrochemical methods of analysis. Special reviews cover voltammetric analysis of waters [18–20]. The use of stripping voltammetry for the investigation of biologically active substances [30, 31] in phase analysis and in the investigation of solids [14, 21, 32] has been reviewed, and so has the use of organic reagents in stripping voltammetry [33] and in situ modified electrodes [34] as well as speciation of the physicochemical forms of metals in solutions [20]. Chemical and biochemical sensors are treated in two volumes of *Sensors* [35]. A strategy for trace metal determination in sea water by anodic stripping voltammetry using a computerized multitime-domain measurement method was described by Bond [36]. However, a simple comparison of the literature cited by Russian and foreign authors shows that there is a language and information barrier: as a rule, most of the Russian papers escape the mention of foreign authors [25, 37]. This book will attempt to fill, to some extent, the information gap and to make recent developments in Russia more accessible to the rest of the world.

At present we can speak about stripping electroanalytical methods as a self-contained branch of analytical chemistry with developed methodology and experimental techniques.

Over the past decade a great variety of new theoretical and experimental developments appeared in the field of stripping electroanalytical methods and some monographs were published, dealing with some aspects of stripping methods. This group of analysis methods was developed so actively that it has become necessary to sum up and to think over the newly accumulated material.

In this book the authors have made an attempt to consider all variants of stripping methods as a set of methods used to study and to analyze not only solutions but also solids. The other goal was to reflect the new qualitative standards attained by currently used stripping electroanalytical methods.

In accord with the task posed, the book presents modern ideas about initial crystallization stages determining concentration conditions in stripping methods, about interaction of components in concentrates, and about electrodeposition of elements initiated by other elements; it gives estimates of the major sources of systematic and random errors occurring in stripping methods and describes possible ways in which the errors can be taken into account or eliminated (Chapters 1 and 2).

The latest achievements in the field of electrode materials, design of electrodes and electrochemical cells used in stripping methods have been critically assessed in Chapter 3.

The authors formulated possible general approaches to the choice of rational schemes of analyzing solutions by stripping methods as applied to various industrial and environmental materials. Optimum conditions for the determination of more than 40 elements of the Periodic System in these materials are given in the form of tables and diagrams (Chapter 4).

The use of stripping methods in solid state chemistry is reflected in Chapters 5 and 6, which are dedicated to the search for electrochemical analytical responses to particular specific features of the solid structure.

A limited volume of the book prevented us from giving a complete list of references to numerous original papers and therefore we had to restrict ourselves in many cases to the corresponding reviews citing those papers. We apologize to the authors concerned.

The Introduction and Section 2.1 were written by Kh. Z. Brainina and E. Y. Neyman; the Preface, Chapters 1, 3, 5, 6 and Sections 2.2, 2.4, and 2.5 were written by Kh. Z. Brainina, Section 2.3 and Chapter 4 by E. Y. Neyman.

The authors express their sincere and deep appreciation to academicians I. P. Alimarin and Y. A. Zolotov for helpful recommendations regarding the analytical part of the work; Prof. V. M. Zhukovski and Dr. M. Y. Hodos for discussions about Chapter 6; A. V. Tchernysheva, N. Y. Stozhko, L. N. Kalnishevskaya, R. M. Khanina, M. B. Vidrevich, G. M. Belysheva, and V. I. Ignatov for their participation in the experiments and help rendered in preparing the manuscript for publication.

The authors are indebted in advance to all the readers who will express their opinion on strong and weak points of the book.

KH. BRAININA
E. NEYMAN

Ekaterinburg, Moscow

CONTENTS

ABBREVIATIONS

Methods

AAS	Atomic absorption spectrometry
a.c. SV	Alternating current SV
AdSV	Adsorptive SV
ASV	Anodic SV
CSV	Cathodic SV
d.c. SV	Direct current SV
DPSV	Differential pulse SV
LSSV	SV with linear potential scan
NPSV	Normal pulse SV
PSA	Potentiometric stripping analysis
SCA	Stripping chronoamperometry
SCP	Stripping chronopotentiometry
SEAM	Stripping electroanalytical methods
SV	Stripping voltammetry
SWSV	Square wave SV

Electrodes

CE	Carbon electrode
CPE	Carbon paste electrode
CPEE	Carbon paste electroactive electrode (containing the analyte substance)
GCE	Glassy carbon electrode
GE	Graphite electrode
GPE	Graphite paste electrode
Hg/CE	Mercury plated carbon electrode
HMDE	Hanging mercury drop electrode
IGE	Impregnated graphite electrode
M/CE	Metal plated carbon (glassy carbon) electrode
MFE	Mercury film electrode
RDGCE	Ring-disk glassy carbon electrode
SMDE	Static mercury drop electrode

NOMENCLATURE

a	Activity of deposited component, mol/cm^3
a_∞	Activity of component in standard state, mol/cm^3
A	Nucleation rate at an active site, s^{-1}
c^0	Volume concentration of ions, mol/cm^3, mol/L, g/L
c	Content of component in a sample, %
c_{min}	Detection limit, mol/L, g/L, %
c_{ox}	Oxidant concentration, mol/L, g/L
c_s	Metal ion concentration at the electrode surface, mol/cm^3
d	Atom diameter, nm
D	Diffusion coefficient, cm^2/s
E^0, E_1^0	Standard electrode potentials, V
E_p	Maximum current potential, V
E_i	Initial scan potential, V
E_{eq}	Equilibrium potential, V
E_{el}	Preelectrolysis potential, V
ΔE	Potential jump in chronopotentiogram, V
F	Faraday number
ΔG^0	Gibbs energy change, kJ/mol
I	Maximum current, A
I_I, I_{II}	Maximum current of metal oxidation from the first and second energy states, respectively, A
i	Current, A
i_I, i_{II}	Discharge or ionization current of metal in the first and second energy state, A
k_s	Electrode process rate constant at standard potential, cm/s
n	Number of electrons participating in electrode process
q	Quantity of electricity equivalent to the amount of metal (compound) on electrode surface, C
q_I, q_{II}	Quantity of electricity equivalent to the amount of metal M_I and M_{II} on the electrode, respectively, C
q'	Quantity of electricity equivalent to the amount of metal when $q_{II} \to 0$, C
q_s	Quantity of electricity equivalent to metal monolayer on the surface S_1^0, C

Q	Total quantity of electricity, C
S, S_1^0	Total and active surface of the electrode, cm^2
s_r	Relative standard deviation
t	Anodic process time, s
t_{max}	Time required to attain maximum current, s
v	Potential scan rate, V/s: V/min
α, β	Coefficients of cathodic and anodic electrode process transfer, respectively
γ	Proportionality factor reflecting for metal-electrode system peculiarities, 1/C
δ_c, δ_a	Effective thickness of diffusion layer at the cathodic and anodic stages of electrode process, cm
μ	System chemical potential, J
τ_{el}	Preconcentration time, s, min
τ, τ_t	Process transient time, s, min
θ	Fraction of electrode surface occupied by deposit
ω_a, ω_c	Disk electrode rotation speed at the anodic and cathodic stage of electrode process, cm^{-1}
Ω	Maximum concentration of metal on the active electrode surface, mol/cm^3

CHEMICAL ANALYSIS

A SERIES OF MONOGRAPHS ON
ANALYTICAL CHEMISTRY AND ITS APPLICATIONS

J. D. Winefordner, *Series Editor*

Electroanalytical Stripping Methods

INTRODUCTION

Electroanalytical methods are widely used in scientific studies and monitoring of industrial materials and the environment. One of the most widespread electroanalytical methods is voltammetry (polarography). The method permits the determination of a great number of ions of metals and organic compounds with a detection limit 10^{-6}–10^{-5} M.

The idea expressed by Barker in 1952 on the possibility of preconcentrating analytes on the electrode in order to lower detection limits proved extremely fruitful. This approach was one of the main reasons for the revival of polarography and served as the starting point of a new direction called stripping electroanalytical methods (SEAM).

Initially, SEAM were used for the analysis of solutions in two versions which differed in the aggregative state of the concentrate on the electrode: SEAM of amalgams and SEAM of solid phases. In the last decade we have witnessed the development of SEAM used for the investigation of solids. In this connection, it is expedient, to refine the terminology used for SEAM. It would probably, be wise to distinguish between the SEAM used for solutions, including all variants of the SEAM irrespective of the electrode type and concentrate nature, and the SEAM used for solids, where the subject of study and analysis is immediately a solid.

In almost all cases (except phase and elementary analysis of solids) the first stage in SEAM is the accumulation of the electroactive substance on the surface or in the bulk of the electrode. The second stage is informative and is conducted under various modes of electrode polarization: potentiodynamic conditions, including superposition of pulses of different shape and amplitude and galvanostatic and potentiostatic conditions. Correspondingly, the information sources in SEAM are cathodic or anodic voltammograms (stripping voltammetry), potential-time dependences (stripping chronopotentiometry), or current-time dependences (stripping chronoamperometry and coulometry) (see the diagram). The first method of information extraction is used most widely.

Electrode reactions used in SEAM are given in the diagram as combinations of different stages: electrochemical, chemical, and adsorption.

One of the given stages (reactions) or their combination is used for the

1

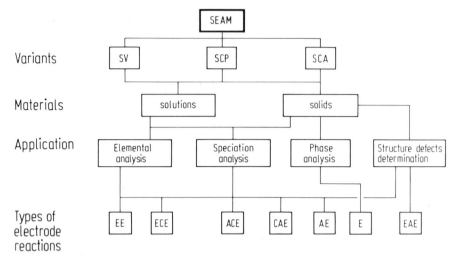

Variants			
Materials			
Application			

E, EE, ECE, ACE, etc. denote the sequence of electrochemical (E), chemical (C), and adsorption (A) stages.

concentration (accumulation) of the electroactive substance. The source of information in SEAM is always an electrochemical reaction.

The great variety of electrode reactions, electrode materials (including solids studied), electrode designs, and conditions of electrode polarization determines a wide field of SEAM application. Essential advantages of SEAM over other methods used to determine trace concentrations of inorganic and organic substances in solutions and to analyze and study solids are:

- The possibility of determining a considerable number (over 40) of chemical elements of the Periodic Table and many organic substances
- Low detection limits attaining 10^{-9}–10^{-10} M for some elements (Cd, Bi, Tl, Pb, Ni) and organic substances
- High selectivity of SEAM, good accuracy and reproducibility
- The possibility of determining correlation of the "composition-property" for the solid substances and materials analyzed and of studying structural peculiarities of these entities
- Ease of computerization and automation of analytical procedures
- Relative simplicity and low price of the instruments used in SEAM

Over the past few years the analysis of traditional metallurgical materials, semiconductor compounds, and high-purity reagents has been relegated to the background, while the leading positions have been occupied by the analysis of environmental materials (water, soil, plants, the atmosphere), biological and clinical materials, and medicines.

In recent decades significant progress has been made in the study of phase and elementary composition, nonstoichiometry, adsorptivity, and the defect structure of solids, based on the use of the electrochemical response of the system as the source of information. It is possible to obtain information by electrochemical methods because the properties of a solid, including the electrochemical properties, depend on the solid's structure (crystal and electronic) and defects (deviation from stoichiometry, presence of impurities, vacancies, interstitial atoms).

We shall consider here the processes at the "solid substance studied–aqueous electrolyte solution" interface, taking as the information signals the current of the electrochemical transformations of solids and of substances adsorbed on their surface.

The prospects for using SEAM in this area are unquestionable and can be hardly assessed now. It should be also expected that SEAM will play an increasing role in tackling physicochemical problems connected with chemisorption and catalysis. The first, rather successful, steps have already been made. Now it is important to improve experimental techniques and to search for correlations between the responses and parameters of solids.

CHAPTER

1

DISCHARGE-IONIZATION OF METALS

1.1. VOLTAMMOGRAMS OF INDIVIDUAL METALS

Experimental studies of the metal discharge-ionization on the surface of indifferent electrodes show that these processes are described by voltammograms with one or several current peaks [14, 38–40].

The voltammogram type depends on the nature of the electrode and the metal deposited on the electrode, the type of metal–electrode interaction, electrolysis conditions (electrode potential, concentration of metal ions in the solution, deposition time), and physical state of the metal (deposit on a foreign surface, powder in the bulk of an indifferent matrix, e.g., in a carbon-paste electrode).

The great variety of voltammogram types in illustrated in Fig. 1.1. It is readily seen that anodic currents can be observed at potentials both more positive and more negative than the equilibrium Nernst potential E_e. Curve *1* corresponds to oxidation of thallium from two energy states (absorbed Tl_I and phase Tl_{II}) and is interpreted well in terms of the well-devised notions [39–41]. Branch A (curve *2*) describes the oxidation of cadmium powder from a carbon-paste electroactive electrode (CPEE), branch B describes the reduction of ions from the near-electrode layer on the electrode surface, and branch C shows oxidation of the electrochemically formed deposit. In the last instance the potential of peak C is more negative than that of peak A. For discharge-ionization of lead on the surface of a germanium electrode, three types of voltammograms (curves *3–5*) without sharp transitions were observed [42] depending on the metal ion concentration: curves whose anodic branch shows more distinct peaks at potentials more negative or more positive than the equilibrium potential, or curves whose ascending or descending branch is distorted. The electronegative peak is registered first: As the concentration of lead ions in the solution rises, the electronegative peak is deformed and is masked with the electropositive peak, the latter becoming subsequently the only one observable. The cyclic curve of the discharge-ionization of lead (curve *6*) on the surface of a germanium electrode exhibits a cathodic-current "loop" after the polarization direction is changed from cathodic to anodic. The metal is oxidized further at potentials more negative

5

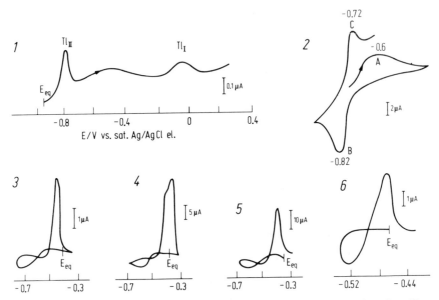

Figure 1.1. Voltammograms of thallium (1) deposited on the graphite electrode surface ($E_{el} = -1.2$ V, $\tau_{el} = 5$ min); of cadmium introduced into a CPE (2); of lead on germanium electrode (3–6). Electrolytes: 0.1 M NaClO$_4$ + 0.01 M HClO$_4$ (1); 0.5 M H$_2$SO$_4$ (2); 0.1 M HCl (3–6). $c^0(Tl^+) = 2 \times 10^{-9}$ M; $c^0(Pb^{2+}) = 6.8 \times 10^{-5}$ (3); 2×10^{-4} (4); 2.6 $\times 10^{-4}$ (5); 3.4 $\times 10^{-5}$ M (6).

than the Nernst potential. Analogous cyclic curves were observed by Fletcher and co-workers [43]. At least two facts must be explained here: appearance of anodic current at potentials more negative than the equilibrium potential and cathodic current hysteresis.

The processes responsible for the appearance of the curves of the first type (curve 1) have been considered in numerous works dealing with undervoltage deposition. As regards curves of type 3–6, we find some isolated considerations. The curves of the type given for cadmium on a CPEE (curve 2) have probably not been described.

Unfortunately, there is no unified mathematical model describing all the given cases. Classical and atomistic models of nucleation, the role played by over- and undervoltage at the stages of electrocrystallization have been considered by Polukarov [44]. The authors [14, 38, 41, 42] trace the relationship between the character of the electrodeposition process (in particular, realization of different energy states of the metal on the surface of a foreign electrode) and the type of metal ionization anodic curve. Let us consider some approaches and, proceeding from these, try to interpret various types of voltammograms and find the solution of the inverse problem, that is, determine the formation conditions of the signals connected with the concentration of particular ions in the solution (analytical signal or response) and obtain information about the energy state of the metal on the electrode surface.

1.1.1. Strong and Weak Interactions—Adatom Formation Conditions

Consideration of electrode processes, including the appearance and disappearance of the solid phase, is incomplete without examination of specific features of this phase.

Data on stripping voltammetry [14, 38, 41], measurements of equilibrium potentials of metals deposited on a foreign electrode surface [14, 38], the electronic work functions depending on the number of monolayers [45], and data on electron diffraction [46] lead one to the conclusion that the deposition process can involve the formation of highly ordered layers without crystal structure, where the energy state of the metal differs, probably, from the energy state in the crystal lattice.

The conclusion on the existence of metal in different energy state is confirmed also by the curves depicted in Fig. 1.1. Most likely, either a continuous or discrete energy spectrum can be realized. At present there is no unified model that would describe the processes of nucleation and disappearance of the solid phase with allowance for the circumstances described above.

Two approaches have taken shape in describing discharge-ionization of metals on solid electrode. One approach is used in electrocrystallization studies [44], where the principal attention is paid to the formation of atoms adsorbed on the electrode surface (adatoms), surface diffusion, implantation of adatoms at the growth sites, dislocations, and so on. The other approach has been worked out in stripping and cyclic voltammetry concerned with nucleation thermodynamics of a new phase [14, 40].

The terminology used in both approaches is ambiguous. Thus, the notion of adatom is understood differently in particular studies. In the theory of electrocrystallization, adatoms are atoms on the like electrode, which have not yet occupied growth sites and which in their totality represent a formation richer in energy than the metal in the lattice volume. In stripping voltammetry [14, 38, 40, 47], adatoms (adsorbate) are thought to be a layer formed on the surface of a foreign electrode in the undervoltage conditions. The Gibbs energy of the layer is probably less than that of the bulk metal. Even within the framework of a single approach the notion of activity, in particular, is used to describe properties of an adsorbate [40] and phase layers [14, 38].

Thus, in our opinion, the problem amounts to consideration of the discharge-ionization of metals on an indifferent electrode as a unitary process that includes formation of adsorption and phase layers, where the type of interaction and the predominant direction are determined by the nature of the metal-electrode system and by electrolysis conditions (concentrations, potentials).

To avoid misunderstanding due to terminology, we introduce the notion of three energy states of the metal $(M_I - M_{III})$ on the electrode surface, which are formed according to the reactions:

$$M^{n+} + ne \quad \begin{array}{l} \nearrow M_I \qquad \Delta G^0 < 0 \qquad\qquad (1.1) \\ \longrightarrow M_{II} \qquad \Delta G^0 \approx 0 \qquad\qquad (1.2) \\ \searrow M_{III} \qquad \Delta G^0 > 0 \qquad\qquad (1.3) \end{array}$$

We assume that the metal M_I in the first energy state (adsorbate, adatoms) is bound to the electrode surface more strongly than in the corresponding crystal lattice (metal in the second energy state M_{II}, metal phase). Nuclei of the metal in the third energy state M_{III} are weakly bound to the electrode surface.

Let us refer to Eq. 1.1 as a strong interaction and Eqs. 1.2 and 1.3 as weak interactions. Note that the terms "strong" and "weak" are used conventionally. We speak only about more or less strong forms of metal binding to the electrode surface, which can be realized in one and the same system depending on particular conditions. The fact that M_I can be formed in principle in the electrode process is determined by the condition $\Delta G^0 < 0$ for the process $M_{II} \rightarrow M_I$ [14, 39]. Using the obvious relations

$$\tilde{\mu}_{M^{n+}} + n\tilde{\mu}_{M^e} = \mu_{II}^0 \qquad\qquad (1.4)$$

$$\tilde{\mu}_{M^{n+}} + n\tilde{\mu}_{E^e} = \mu_{I}^0 \qquad\qquad (1.5)$$

is easy to show that

$$\Delta G^0 = \mu_I^0 - \mu_{II}^0 = n(\tilde{\mu}_{E^e} - \tilde{\mu}_{M^e}) \qquad\qquad (1.6)$$

where μ_I^0 and μ_{II}^0 are chemical potentials of the metal in the first and second energy states; $\tilde{\mu}_{M^{n+}}$, $\tilde{\mu}_{M^e}$, and $\tilde{\mu}_{E^e}$ are electrochemical potentials of the M^{n+} ions and electrons in the metal being deposited and in the electrode material.

The presence of hysteresis on the cathodic branch of the curve and appearance of anodic currents at potentials more negative than the oxidation potentials of the corresponding phases suggest realization of the nonequilibrium state of the metal, M_{III} (Eq. 1.3). Naturally, in this case the following relation should hold:

$$\Delta G^0 = \mu_{III}^0 - \mu_{II}^0 > 0 \qquad (1.7)$$

Strictly speaking, the systems considered are not equilibrium ones. They change with time. However, taking into account the slow rate of the recrystallization and structure ordering processes, the thermodynamic approach may be used most likely for qualitative considerations within small time intervals.

Reactions 1.1–1.3 can proceed either in sequence or in parallel depending on the nature of the metal-electrode system and electrolysis conditions. Processes 1.1 and 1.2 are considered in this section, while Process 1.3 will be treated in the section that follows. It is believed that Process 1.1 proceeds on the active surface S_1 which is characterized by the exponential energy distribution of active sites:

$$S_1 = S[1 - \exp(\Delta G^0/RT)] \qquad (1.8)$$

Metal in the first energy state may cover partially or completely the active surface of the electrode. In this case the relationship $q_1 \leqslant q_s$ must be met

$$q_S = nFS_1 \Omega d \qquad (1.9)$$

The field of the concentrations c of the M^{n+} ions is given by

$$\frac{dc}{dt} = D\frac{d^2c}{dx^2}, \quad 0 < x < \infty, \quad t > 0 \qquad (1.10)$$

with the following initial and boundary conditions:

$$c(x,0) = c_0 \quad \text{when } x \geqslant 0$$
$$c(x,t) = 0 \quad \text{when } x \to \infty \qquad (1.11)$$

$$i_1(t) = nFS_1 k_S \left\{ c(0,t)\left(1 - \frac{q_1(t)}{q_s}\right)\exp\left[-\frac{\alpha nF}{RT}(E - E^0) - \frac{\alpha \Delta G^0}{RT}\right] \right.$$

$$\left. - \frac{q_1(t)}{q_s}\Omega\exp\left[\frac{\beta nF}{RT}(E - E^0) + \frac{\beta \Delta G^0}{RT}\right]\right\}, \qquad (1.12)$$

$$i_{\mathrm{II}}(t) = nFSk_{\mathrm{s}}\left\{ c(0,t)\exp\left[-\frac{\alpha nF}{RT}(E-E^0) \right] \right.$$

$$\left. - a_{\mathrm{II}}(t)\exp\left[\frac{\beta nF}{RT}(E-E^0) \right] \right\}, \tag{1.13}$$

$$i = -nFSD\frac{dc(0,t)}{dx}, \quad \text{where } i = i_{\mathrm{I}} + i_{\mathrm{II}}, \tag{1.14}$$

$$a_{\mathrm{II}} = a_{\infty}(1 - e^{-\gamma q_{II}}). \tag{1.15}$$

A detailed analysis of the solution to the Eqs. 1.10–1.15 is given in [14, 41, 48]. Here we shall consider principally the conditions of formation of M_{I} (adatoms) and their role at the initial stages of electrodeposition.

The calculated dependence of the ratio of the active surface to the entire area of the electrode on the energy gain during deposition of the metal M_{I} is shown in Fig. 1.2. It is seen from the figure that the curve has a portion, where the quantity $S_{\mathrm{I}}^0/S \to 1$ is independent of ΔG^0, that is, the entire electrode surface is active. The calculated dependence $q_{\mathrm{S}} = f(\Delta G^0/RT)$ is of a similar shape.

A change in ΔG^0 as referred to the electrode of the same type testifies to a change in the nature of the metal being deposited. With ΔG^0 being the same, filling of the electrode active surface depends on the electrolysis time and the concentration of metal ions in the solution:

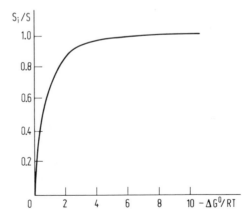

Figure 1.2. Calculated dependence of S_1/S on $\Delta G^0/RT$. Parameters: $c^0 = 1 \times 10^{-9}\,\mathrm{mol\,cm^{-3}}$; $\eta = E - E_{\mathrm{eq}} = -0.4\,\mathrm{V}$; $\tau_{\mathrm{el}} = 5\,\mathrm{min}$; $\alpha = \beta = 0.5$; $\gamma = 10^5\,\mathrm{C^{-1}}$; $D = 10^{-5}\,\mathrm{cm^2/s}$; $\Omega = 0.1\,\mathrm{mol\,cm^{-3}}$; $S = 10^{-2}\,\mathrm{cm^2}$; $n = 1$; $k_{\mathrm{s}} = 10^{-2}\,\mathrm{cm/s}$, $a_{\infty} = 1$.

- In systems with a high energy gain ($\Delta G^0 \ll 0$) the metal is deposited mainly as M_I; that is, in this case, if the electrode potential is close to the equilibrium potential, it can be expected that a monolayer of the deposit is formed.

- In systems with a low energy gain ($\Delta G^0 \approx 0$) the metal is deposited mainly as M_{II} irrespective of the concentration and electrolysis time at potentials more negative than the equilibrium potential.

- In systems with mean values of ΔG^0 the deposit condition is determined by the electrodeposition conditions, since in this case both M_I and M_{II} can be realized.

Depending on ΔG^0 and the concentration, the curves exhibit one or two current maxima. Two current peaks always correspond to dissolution of the metal from different energy states. A voltammogram with one current peak can describe dissolution of the metal both from a single energy state and from two energy states. The higher $|\Delta G^0|$, the more pronounced are both peaks and the broader is the concentration interval over which the peaks show up.

So, the shape of the curves, the number of the signals, and their magnitude and sequence reflect the nature of the deposited element and the electrode, electrolysis conditions, and the character of binding of atoms to the electrode surface and of interatomic bonds.

Calculations and experimental data show that a linear correlation exists between the difference of the peak potentials and the change in the Gibbs energy [14, 39]. Considering the difference of the peak potentials in voltammograms as a measure of the difference between the metal-electrode and metal-metal binding energies and using $\Delta \Phi$ (the difference between the electronic work functions of the metal and the electrode), the authors [39] established a linear correlation of the form

$$E_M = \alpha \Delta \phi \tag{1.16}$$

where $\alpha = 0.5 \text{ V/eV}$.

A similar linear correlation between ΔE_M and $\Delta \phi$ follows from Fig. 1.3. Systems are considered where discharge-ionization processes are not complicated by complex formation and adsorption of halogenide ions. The shape of the curves depends on neither of the parameters used in the theory except γ [38], which is determined from experimental data by extrapolation of the linear portion of the curve $I = f(q_{II})$ to the abscissa axis according to Eqs. 1.17 and 1.18.

$$I = -0.93 \frac{nF}{RT} \beta v q'_2 + 0.835 \frac{nF}{RT} \beta v q_2^0 \tag{1.17}$$

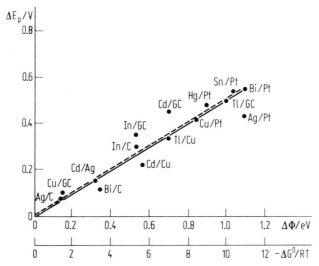

Figure 1.3. Differences in peak potentials ΔE_p on experimental voltammograms as a function of the corresponding electronic work functions $\Delta\phi$ (solid line) and ΔE_p on calculated voltammograms as a function of $\Delta G^0/RT$ (dashed line) (GC, glassy carbon). Parameters; $k_s = 10^{-2}$ cm/s; $\Omega = 0.1$ mol/cm^3; $\alpha = \beta = 0.5$; $\gamma = 2 \times 10^4\,\text{C}^{-1}$; $D = 10^{-5}$ cm^2/s; $S = 10^{-2}$ cm^2; $c^0 = 1 \times 10^{-9}$ mol/cm^3; $E - E_{eq} = 0.4$ V; $d = 10^{-7}$ cm; $\delta = 10^{-3}$ cm; $n = 1$; $a_\infty = 1$; $v = 0.1$ V/s.

$$I = -0.93\frac{nF}{RT}vq_2' + 0.93\frac{nF}{RT}vq_2^0 \tag{1.18}$$

which hold at $\gamma q_2^0 \gg 1$ for the irreversible and reversible processes, respectively; q_2' is an x-intercept on extrapolation to $I = 0$ of the dependence $I - q_2^0$.

As is seen, the dependences given in Fig. 1.3 are practically coincident and are directly proportional. Then a simple relation follows

$$\Delta G^0 = \alpha k \Delta\phi \tag{1.19}$$

where $k = 8.0658\,\text{eV}/8.359\,\text{J·mol}$ serves for correlating the systems of units.

Thus, the value of ΔG^0 required to calculate voltammograms corresponding to a certain metal-electrode system can be found using Eq. 1.19 and tabulated values of the electronic work functions. At the same time, ΔG^0 can be determined from experimental voltammograms using the set of calculated data depicted in Fig. 1.3.

We are aware that the limitations of this approach, result from disregarding the presence of the double electric layer, hydration effects, and different ϕ of the monolayers and the corresponding phase. We believe nevertheless that it is reasonable to exploit these concepts for the interpretation of experimental data and evaluation of ΔG^0 that are necessary for calculations and probabilistic prediction of the behavior of the initial stages of electrodeposition of elements on a foreign substrate.

Let us arrange the elements which are subject of investigation in stripping electroanalytical chemistry in decreasing order of the electronic work function ϕ (for the most reliable values of ϕ see [49]):

$$Pt, As, Pd, Rh, Os, Ru, Au, Se, Te, C, Cu, Sb, Hg,$$

$$Sn, Bi, Ag, Zn, Ga, Pb, In, Cd, Tl \qquad (1.20)$$

Note that the potentials of the zero charge of metals and the ability of metals to orient water dipoles change following practically the same order. Hence, the effects of the double layer and hydration do not influence significantly the position occupied by the metal in the given series.

Let us consider electrolysis of a solution containing an indifferent electrolyte and electroactive ions M^{n+} on a solid electrode E which is foreign to the metal M. The following cases are possible:

1. $\phi_M < \phi_E$, $\Delta\phi_{M-E} < 0$
2. $\phi_M \approx \phi_E$, $\Delta\phi_{M-E} \approx 0$
3. $\phi_M > \phi_E$, $\Delta\phi_{M-E} > 0$

Use a carbon (graphite) electrode as the foreign one. Then in the first case it is energetically favorable that elements standing to the right of carbon are deposited on the electrode ($\Delta G^0 < 0$), and they are capable of the undervoltage deposition from low-concentration solutions. As the absolute value of the difference ($\Delta\phi_{M-E}$) of the electronic work functions of the deposited metal and the electrode increases, the energy of interaction of metals with the foreign surface rises and strong metal–electrode interactions show up. Metal in the M_I state is deposited predominantly.

In the second case the energy gain obtained on deposition of the element on the electrode is insignificant ($\Delta G^0 \approx 0$). A weak metal–electrode interaction is observed. Metal can be deposited on the electrode from solutions containing a relative high concentration of electroactive ions. Metal is deposited mainly as M_{II}. Its properties are characterized by the activity a_{II} [14, 38], assuming that it depends on the amount of the deposited metal. Properties of the deposit on a foreign substrate differ from those of the corresponding metal in its own lattice, because of the influence of the electrode structure distorting the lattice of the initial layers of the deposit being formed. In accord with the terminology adopted in [14, 38], the macrophase will imply the metal that possesses properties of a three-dimensional metal of activity a_∞, and the microphase will stand for the deposit whose properties differ from those of the macrophase.

When a graphite electrode is used, the third case is observed for the elements that stand in the series (20) to the left of carbon. For these elements deposition on a graphite electrode is energetically unfavorable ($\Delta G^0 > 0$). As will be shown below, the deposition is sometimes possible only in the presence of some other elements. Cathodic overvoltage is typical of the second and third cases. As a rule, the metal M_{III} is deposited.

1.1.2. Electrocrystallization Overvoltage

It is known that nucleation takes place at a limited number of "active" sites on the surface. Markov and Kashchiev [50] hold to the opinion that the number of such sites depends on the electrode potential. Fletcher and co-workers [43, 51] think that the active site distribution function is characteristic of the electrode surface. The difference is that in the former case the activity of all nucleation-favorable sites is the same at the given overvoltage, while in the latter case [43, 51] a hierarchy in the activity of such sites is suggested. Since experimental data are evidently insufficient for choosing this or that model, Mostany and co-workers [47, 52] used the representations of Markov and Kashchiev [50] as these are simpler.

Then the following relation [47] is true:

$$N_{0,t} = N_0 \exp(-At) \tag{1.21}$$

where $N_{0,t}$ (cm^{-2}) is the number of active sites not occupied by nuclei by the time t; N_0 is the total number of active sites; A is the rate of of nucleation at an active site (s^{-1}).

A decrease in the rate of nucleation on the electrode as a result of the active site transformation into growing nuclei is described by the expression

$$dN/dt = AN_{0,t} = AN_0 \exp(-At) \tag{1.22}$$

Subsequent decrease in the number of active sites is due to their overlap with growing nuclei and diffusion zones. As a rule, the latter factor is decisive.

The expression for the current strength at a given potential, which is derived allowing for this circumstance and Eq. 1.22, has the form

$$i = (zFD^{1/2}c^0/\pi^{1/2}t^{1/2})\left\{1 - \exp\left[-N_0\pi kD\left(t - \frac{1 - e^{-At}}{A}\right)\right]\right\} \tag{1.23}$$

Equation 1.23 has a maximum.

Equation 1.24 determines the time during which the maximum is attained and Eq. 1.25 determines the maximum current

$$\ln\left[1 + 2bt_m - 2bt_m \exp(-At_m)\right] - bt_m + bA^{-1}[1 - \exp(-At_m)] = 0 \quad (1.24)$$

$$I_m = (a/t_m^{1/2})\{1 - \exp[-x + \alpha(1 - e^{-x/\alpha})]\} \quad (1.25)$$

where $a = zF\pi^{-1/2}D^{1/2}c^0$. $(As^{1/2}/cm^2)$; $x = bt_m$ is the dimensionless time corresponding to the current maximum; $b = N_0\pi kD(s^{-1})$; $k = (8\pi c^0 M/\rho)^{1/2}$ is the dimensionless constant characterizing the rate of growth of diffusion zones; M is the molar mass of the deposit (g/mol); ρ is the deposit density (g/cm^3); c^0 is the concentration (mol/cm^3); $\alpha = b/A$ is the dimensionless parameter.

Equations 1.24 and 1.25 make it possible to obtain information necessary for the evaluation of the number of active sites on the surface and the rate of nucleation at an active site from single-stage potentiostatic measurements. Computations suggest that the instant and progressive nucleations represent particular cases of the situation considered by Scharifker and Mostany [47]. These particular cases are described asymptotically using solutions for small and large values of the dimensionless parameter α.

Calculations [47] and the experimental study [52] of the electrodeposition of lead on glassy carbon over a wide range of lead ion concentration in the solution and overvoltages show that the number of active nucleation sites is independent of the concentration of lead ions; the surface density of active sites is low even at high overvoltages. Moreover, the number of active nucleation sites is less than the number of lead adsorption sites on glassy carbon. Indeed, electrodeposition of lead on glassy carbon is observed both at undervoltage [53] and at small overvoltage [51]. The concentration dependence of the nucleation rate is indicative of the fact that atoms are attached to the growing nucleus immediately from the solution volume. As the concentration of lead ions in the solution increases, the nucleation mechanism is changed: progressive nucleation is replaced by instant nucleation.

An increase in the cathodic current was revealed after the scan direction was reversed from cathodic to anodic under conditions of the triangular potential scan for discharge ionization of lead [51] and copper [54] on a glassy-carbon electrode and of lead on a germanium electrode [42]. A loop (see Fig. 1.1) or a current maximum appears on the cathodic branch of the curve. It is natural to attribute a priori this shape of the curve to the manifestation of the crystallization overvoltage which lowers after the first nuclei are formed. Calculations made by Fletcher [51] support these suggestions.

If the rate A of the crystal appearance and the radial rate $k(t)$ of the crystal growth are positive monotonically increasing functions of the single variable E [51], the expression for the current in the case of slow crystallization at

$$E = \begin{cases} vt & 0 < t \leqslant \theta \\ v(2\theta - t) & \theta < t \leqslant 2\theta \\ 0 & \text{other cases} \end{cases}$$

has the form

$$i(t) = nF\rho_M V'(t) \tag{1.26}$$

where θ is the duration of the cathodic stage (s); ρ_M is the mole density of the deposited crystals (mol/m^3); $V'(t)$ is the variation rate of the total volume of all crystals (m^3/s).

The quantity $V(t)$ is preset by the Volterra equation of the first type:

$$V(t) = \frac{2\pi}{3} \int_0^t A(\tau)[R(\tau, t)]^3 dt \tag{1.27}$$

$$R(\tau, t) = \int_\tau^t k(t) dt \tag{1.28}$$

and provided $k(t)$ is not a Function of τ then the first and second derivatives of Eq. 1.27 with respect to time are

$$V'(t) = 2\pi k(t) \int_0^t A(\tau)[R(\tau, t)]^2 dt$$

and

$$V''(t) = 4\pi[k(t)]^2 \int_0^t A(\tau)R(\tau, t) d\tau + 2\pi k'(t) \int_0^t A(\tau)[R(\tau, t)]^2 dt \tag{1.29}$$

where τ is the nucleation time.

Equation 1.26 has a maximum over the interval $\theta < t < 2\theta$. In the case of slow (e.g., semispherical) diffusion, Eq. 1.27 and 1.28 become more complicated, since the crystal growth rate is also a function of the nucleation time. The solution of the problem describing the electrocrystallization process under conditions of slow diffusion is given by Fletcher [51]. We shall not present this solution here and restrict ourselves to noting that in this case, too, a current maximum and a loop appear on the curve $i = f(E)$ over the interval $\theta < t \leqslant 2\theta$, irrespective of the nucleation rule.

A loop on the cyclic curve was observed for discharge-ionization of Zn, Tl, Cd, Pb, Sn, Bi, Hg, Ag, Co, Ni, Cr, and Mn and for deposition of α-PbO_2 and β-PbO_2 on glassy carbon (the curves are analogous to curve 6, Fig. 1.1), [51].

Of great importance is the fact that these results are valid only for nucleation and growth of independent crystals, that is, only for initial electrocrystallization stages, where intercrystalline interactions and overlapping of diffusion zones are improbable.

Thus, the appearance of the current maximum on the current versus time curve during potentiostatic deposition [47, 52] and the loop on the cyclic curve registered during triangular voltage scan [51] can be used as the diagnosis criteria of the fact that the limiting stage of the process is crystallization (nucleation and nuclei growth kinetics), since other kinds of overvoltage (charge transfer, adsorption, diffusion) do not manifest themselves in this way. The only exception can be the rare case of nondiffusion-controlled autocatalysis.

The shape of cyclic curves depends on the electrode polarization potential interval (potential at which the scan direction is changed), time of exposure to this potential, and the concentration of metal ions in the solution. For example, a stop of the potential scan in the region of the metal ion discharge is accompanied by an increase and redistribution of currents on the anodic branch of the curve (cf. curve 6 in Figs. 1.1 and 1.4). With sufficient holding time, the loop disappears on the cyclic curve and only one peak of anodic current is observed at a higher electrode potential. Analogous changes in the shape of the cyclic curve are observed with increasing concentration of lead ions in the solution (curves 3–5, Fig. 1.1). Thus, when the concentration of metal ions in the solution is low, an anodic peak appears at a potential more negative than or close to the equilibrium potential; then the currents suffer redistribution: a peak appears at a more positive potential, becoming predominant as the concentration is raised further.

The influence of the deposition potential on the character of anodic curves is illustrated in Fig. 1.5 showing voltammograms of oxidation of cadmium deposited on cadmium selenide [55]. If the deposition potential is shifted relative to the equilibrium value to the negative side for not more than 0.17 V, anodic voltammograms exhibit one peak (curves 1 and 2, Fig. 1.5) whose magnitude is enhanced as the deposition time is increased. Therewith the current maximum potential is shifted somewhat toward positive values. Ionization of cadmium ($50 < q < 7500\,\mu C\,cm^{-2}$) takes place over a narrow potential interval. The current drops sharply after the maximum and the points of intersection of the ascending branches of the curves and the zero

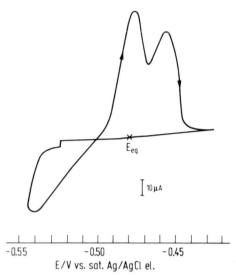

Figure 1.4. Cyclic voltammogram recorded when the scan was stopped for 10 s at a potential -0.525 V, $v = 0.02$ V.s. Solution composition: 0.1 M HCl + 2.5 × 10^{-5} M Pb^{2+}. Electrode $p = Ge$ (conductivity 0.3 Ω·cm) with a hydride coating.

current lines are practically independent of q. These peculiarities are typical of the macrophase.

If rather small amounts of cadmium (equivalent to the values of q less than 10–$15\,\mu C\,cm^{-2}$) are deposited on the electrode at the given potentials, the shape of anodic voltammograms changes are compared to the one described above: the drop of the ionization current after the maximum becomes much smoother (curve 3, Fig. 1.5), this being characteristic of the metal microphase [38]. Different properties of the micro- and macrophases show up also in the presence of two straight portions with different slopes on the plot depicting I_m vs. q. The value of the ratio $(dI_m/dq)_{macro}/(dI_m/dq)_{micro}$ is close to the theoretical value for the reversible ionization process (3.2) [38]. Following the concepts worked out in [38], it is natural to conclude that curves 1–3 (Fig. 1.5) describe oxidation of the metal from its second energy state that includes micro- and macrophases.

In going over to more negative potentials, when $|E - E_{eq}| > 0.18 - 0.20$ V [if $C_{Cd^{2+}} \sim (1$–$4)\cdot 10^{-4}$ M], the ascending branch of the anodic voltammogram shifts toward more negative potentials and becomes increasingly convex (curve 4, Fig. 1.5); subsequently, this convexity transforms to the second (electronegative) anodic peak which becomes predominant (curves 5 and 6, Fig. 1.5) at E_{el} lower than -0.8 V. This suggests that under the given

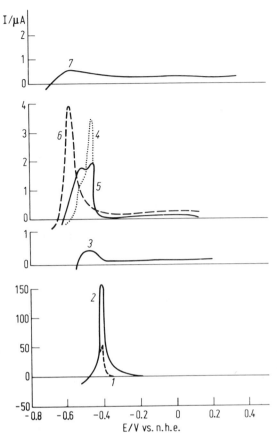

Figure 1.5. Dark anodic voltammograms of ionization of cadmium deposited under potentiostatic conditions from uniformly stirred electrolytes on CdSe electrodes treated preliminarily according to the standard procedure (*1–6*) and a characteristic voltammogram registered in supporting electrolyte (*7*). S, cm^2: *1, 2, 4–7* -0.25; *3* -0.11; v, $V/\bar{S}$: *1, 2, 4–7* -0.02; *3* -0.05; $c_{Cd^{2+}}^0$, g-ion/1: *1, 2* $- 3.2 \times 10^{-4}$; *3* $- 1.1 \times 10^{-4}$; *4–6* $- 4.3 \times 10^{-4}$; *7* $- 0$. Curves *1, 2*: $E_{el} = -0.59$ V; τ, s: *1* $- 129$; *2* $- 435$; $q > 50\,\mu C\,cm^{-2}$. Curve *3*: $E_{el} = -0.61$ V; $\tau = 5$ s; $q < 15\,\mu C\,cm^{-2}$. Curves *4–6*: $\tau = 10$ s; E_{el}, V: *4* $- 0.75$; *5* $- 0.78$; *6* $- 0.91$. Curve *7*: $E_{el} = -0.91$ V; $\tau = 50$ s.

conditions a form of cadmium is obtained, which is more energy rich than the stable modification of the element (third energy state).

Of great significance is a study by Krebbs and Roe [56] who investigated discharge-ionization of silver on single silver crystals. The authors found a higher reactivity of a fresh deposit (0.3–15 at. layers) compared to a three-dimensional specimen. They observed an extremely interesting fact: clearly pronounced peaks of dissolution of this deposit from the surface of the like electrode, a situation which is possible only if the deposit is oxidized at potentials more negative than the base metal. A similar phenomenon is probably responsible for the shift of the potential of the oxidation current maximum of cadmium (electrochemically deposited on the electrode surface) toward negative values as compared to the oxidation current maximum of powdered metal from a CPEE (see Fig. 1.1, curve 2, peaks A and C). Calculations [38, 41] show that under certain conditions the concentration gradient of the dissolving metal ions near the electrode surface can change sign before the Nernst potential is attained. However, this cannot account for the current redistribution and the cathodic current hysteresis (see Fig. 1.1, curves 2–6, and Fig. 1.4) with no consideration to the concepts of slow crystallization and the third energy state.

Formation of the new phase is connected with transition through an intermediate microheterogeneous state that possesses a higher Gibbs energy compared to the initial and final states. Mamaev [57] supposes that the dependence ΔG^0 passes through the maximum when $r = r_{cr}$ (r is the crystal radius). Crystals of critical dimensions exhibit a minimum stability. The excess energy is necessary to set up a host of interfaces between the previous phase and nuclei of the new phase whose surface energy is high. In this third energy state the metal can probably be represented by small thermodynamically unstable and energy-rich nuclei and clusters. Naturally, their formation requires some overvoltage (Gibbs–Thomson effect). A decrease in the cathodic process overvoltage subsequent to deposition of some amount of metal on the electrode surface manifests itself in a considerable growth of the cathodic current after the potential scan direction is reversed from cathodic to anodic when recording a cyclic curve. This accounts for the appearance of a loop. Oxidation of such energy-rich nuclei proceeds, naturally, at a lower anodic polarization than the oxidation of the corresponding equilibrium phase. An increase in the amount of metal on the electrode surface with increasing concentration of the corresponding ions in the solution, shifting of the metal deposition potential toward negative values, and longer duration of the process leads to enlargement of nuclei and to their overlap; the contribution of the surface energy to the system's energy diminishes, resulting in shifting of the dissolution potentials toward the equilibrium and more positive values, in the redistribution of currents in favor of the electropositive peak, and in a smaller loop. Thus, in [51] no oxidation peaks were observed at potentials equal to or more negative than the equilibrium potential, because by the time of dissolution the crystal dimension was sufficiently large (about 1 μm).

1.1.3. Criteria for the Realization of Different Energy States of Metals

The energy spectrum of metals can be presented as follows:

$$
\begin{array}{ccc}
M_I & M_{II} & M_{III} \\
\Delta G^0 < 0 & \Delta G^0 \approx 0 & \Delta G^0 > 0 \\
\eta < 0 & \eta \approx 0 & \eta > 0
\end{array}
$$

$$\eta = E_{eq} - E_m$$

Note that the position of the anodic current peak relative to the equilibrium potential does not always permit drawing an unambiguous conclusion on the energy state of the dissolving metal. Calculations made using Eqs. 1.10–1.15 for the triangular potential scan show that even when $\Delta G^0 < 0$ the anodic current peak can be observed at a potential more negative than the Nernst potential [58].

Thus, for $\Delta G^0 = -RT$,

$$E_M \approx E' > E_{eq} \quad \text{when} \quad q_I > q_{II}$$

$$E_M \approx E' < E_{eq} \quad \text{when} \quad q_I < q_{II}$$

for $\Delta G^0 = -0.2RT$,

$$E_M > E_{eq}; \quad E_M > E' \quad \text{when} \quad q_I > q_{II}$$

$$E_M \approx E_{eq} > E' \quad \text{when} \quad q_I \approx q_{II}; \quad E' < E_M < E_{eq} \quad \text{when} \quad q_I < q_{II}$$

where E' is the potential at the moment when $c(0, t) = c^0$. The difference of E' from the calculated Nernst potential occurs because in the latter case the solid phase activity is taken to be equal to 1. The following parameters were used in calculations: $k_S = 10^{-2}$ cm/s; $\alpha = \beta = 0.5$; $\gamma = 10\,C^{-1}$; $D = 10^{-5}$ cm^2/s; $\Omega = 0.05$ mol/cm^3; $E_1 = E^0$; $d = 7 \times 10^{-7}$ cm; $\delta = 3.6 \times 10^{-3}$ cm; $n = 2$; $a_\infty = 1$; $v = 0.04$ V/s; $E_2 = 0.54$ V (E_1 and E_2 being the initial and final potentials); $7 \times 10^{-7} \leqslant c^0 < 5 \times 10^{-4}$ M.

It is also important that the curve with one current peak can describe oxidation of the metal M_I and M_{II}, this being typical of systems with small negative ΔG^0 (small difference between the electronic work functions of the metal and the electrode). Only the presence of two (or three) peaks, one of which is close to the equilibrium value and the other is observed at a more positive potential, presents sufficient evidence (for stageless oxidation) that the metal exists on the electrode in the first and second energy states. An additional criterion, which is not always fulfilled, is saturation of the dependences $I_I - c^0$ and $I_I - \tau$.

Appearance of an anodic current peak at a potential more negative than the equilibrium potential accompanied by the emergence of a current peak at a more positive potential as the concentration and deposition time are increased, redistribution of currents in favor of the latter peak, and the presence of the cathodic current loop testify to the realization of the third state of the metal M_{III}.

Thus, depending on the nature of the metal–electrode system and the conditions of electrolysis on a foreign surface, the metal can be formed in three energy states: (i) minimum-energy state; (ii) equilibrium state; (iii) energy-rich state. The first and the third states may be considered as limiting states. In the first case, one should probably expect the formation of two-dimensional nuclei and layer-by-layer uniform growth of the film. Markov and co-workers [59] indicate that in the last case a critical oversaturation takes place, which is energetically favorable for the formation of three-dimensional nuclei, a circumstance which results subsequently in an islet-pattern growth of the deposit. It is natural that in actual systems superposition of different states is possible. For example, Fletcher [51] found lowering of the deposition overvoltage of metal on glassy carbon in the series $Hg > Cu > Cd > Sn > Zn > Tl$, Pb coinciding with the series (1.20). The crystallization overvoltage lowers as the interaction of the metal with the electrode surface increases (the difference between the electronic work functions and the Gibbs energy increases), a situation where deposition of the metal M_1 becomes more probable. This infers that the discussion of the mechanism of the initial stages of nucleation [60] should take into account the possibility of the monoatomic layer formation preceding the stage of the metal phase nucleation proper.

Many systems do not exhibit three-dimensional nucleation. This tendency should naturally increase in the series given above. The knowledge of the conditions causing the superiority of the M_I, M_{II}, and M_{III} metal states is necessary, in particular, for choosing the appropriate analytical signal and conditions of the signal generation.

1.2. VOLTAMMOGRAMS OF BINARY METAL SYSTEMS

The electrochemical behavior of a metal system depends on various factor. Information about these factors is available in phase diagrams.

In some papers [39, 49], the authors persistently keep to the idea that the electronic work function ϕ can serve as an important quantitative characteristic of the electrochemical behavior of a metal. The usefulness of this quantity for describing and predicting interactions in the metal–electrode system is shown in [14, 39, 41], which are expanded on in Section 1.1. Moreover, a fact of simple correlations between ϕ and different properties of metals is known [49]. In particular, an agreement between the "composition-electronic work function" diagrams and the phase diagrams of metal systems has been established [61].

Let us compare the electronic work function of metals and the interaction type (phase diagrams) of binary systems.

Let us arrange selected metals in the descending order of ϕ horizontally and vertically (Table 1.1) and enter general data on the character of the phase diagrams in each box that corresponds conventionally to the system $E_{(1)}-E_{(2)}$. The table specifies groups of elements forming: (i) continuous solid solutions;

Table 1.1. Work Functions, Kinds of Interaction in Binary Systems and Types of Anodic Voltammograms of Electrochemically Formed Deposits

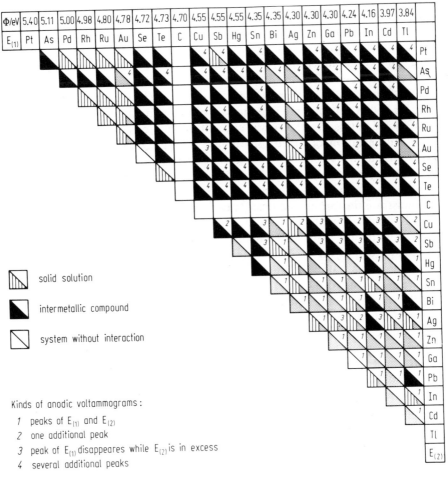

Kinds of anodic voltammograms:

1 peaks of $E_{(1)}$ and $E_{(2)}$
2 one additional peak
3 peak of $E_{(1)}$ disappeares while $E_{(2)}$ is in excess
4 several additional peaks

Legend:
- solid solution
- intermetallic compound
- system without interaction

(ii) intermediate phases and chemical compounds; (iii) eutectic and peritectic alloy systems. Platinum metals exhibiting high values of ϕ (5.4–4.7 eV) are characterized by the similarity of metallochemical properties (in particular, of lattice parameters) and, as a result, by the formation of a continuous series of solid solutions. Phase diagrams with intermediate phases and chemical compounds that are formed along with limited solid solutions are most typical of binary alloys of noble and other metals. The values of the electronic work function of $E_{(1)}–E_{(2)}$ pairs vary considerably. As $\Delta\phi$ increases, formation of compounds in the system becomes more probable.

According to the phase diagrams, metals such as Sn, Pb, Cd, Zn, Tl, Ga, are least prone to the formation of compounds with each other. They are characterized by limited solubility and formation of eutectic and peritectic structures. Note that metals of this group feature the lowest value of the electronic work function. Hence, it may be assumed with a certain degree of reliability that for some binary metal systems the character of interaction and properties of the phases being formed are determined by the values of the electronic work function of the corresponding metals. The systems, where continuous solid solutions are formed, are characterized by anodic curves of the first and second types. Oxidation of electrochemical concentrates that are formed in systems with intermediate phases and intermetallic compounds (IMC) is described by the curves of the third and fourth types depending on the concentration of metal ions in the solution. Curves of the first type correspond to eutectic and peritectic systems. Some exceptions to these regularities are likely due to the difference in the energy state of the metal in the equilibrium alloy matrix and in the electrochemical concentrate. Let us illustrate the formulated postulates using some examples. In doing so, we shall compare anodic potentiodynamic curves of equilibrium alloys and metals with those of the metals deposited from solutions. Use the voltammetry technique employing a CPEE containing powder of the metal or alloy studied. Record anodic–cathodic–anodic curves. Figure 1.6 depicts such a sesquicyclic curve of the Cu–Ni alloy. In this case the first anodic curve describes oxidation of the starting sample, the cathodic curve 2 represents reduction of the metal ions formed in the near-electrode layer as a result of the sample oxidation, and the second anodic curve 3 describes oxidation of the deposit appearing on the electrode at the cathodic stage. To solve the problem posed, one can also use stripping voltammetry of the solution containing ions of the corresponding metals in certain concentrations.

In the copper–nickel system, a continuous series of solid solutions is formed. Electrooxidation of the alloys is given by curves with one current maximum located on the potential axis between the potentials corresponding to oxidation of the individual alloy components. The dependences of the anodic current maxima and the electronic work functions on the alloy com-

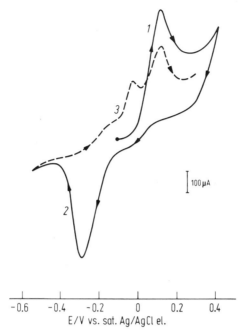

-0.6   -0.4   -0.2   0   0.2   0.4

E/V vs. sat. Ag/AgCl el.

Figure 1.6. Anodic–cathodic voltammograms of Cu–Ni alloy introduced into CPEE (*1, 2*) and anodic voltammogram of the metals deposited on the electrode surface (*3*). 0.5 M KCl, $v = 80$ mV/s, 40% of Ni.

position are analogous (Fig. 1.7). From this it can be inferred that the voltammograms reflect the electronic structure of the Cu–Ni alloys with varying composition and correlate well with the phàse diagram of the system.

It is interesting to note that oxidation of the deposit produced on the electrode surface over the potential interval corresponding to the limiting diffusion current of both alloy components is described by a curve with one peak. Oxidation of the deposit produced on the electrode surface under potentiodynamics conditions is described by curves with two current peaks of individual metals.

The phase diagram of the indium–antimony system can be viewed as a combination of two simpler eutectic diagrams on the basis of In + InSb and InSb + Sb. The composition dependence of the electronic work function changes sharply [62] its character over the concentration intervals, where intermediate phases or chemical compounds exist.

An examination of the cyclic curves in Fig. 1.8 shows that dissolution of the equilibrium InSb alloy (curve *1*) is accompanied by simultaneous ionization of both components (voltammogram with one peak). The metals are deposited

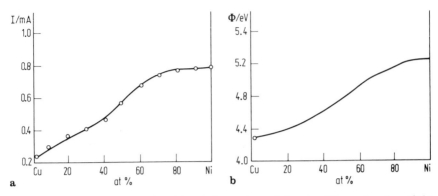

Figure 1.7. Anodic current maxima (*a*) and electronic work function (*b*) as a function of the Cu–Ni alloy composition 0.5 M KCl; $v = 80$ m V/s.

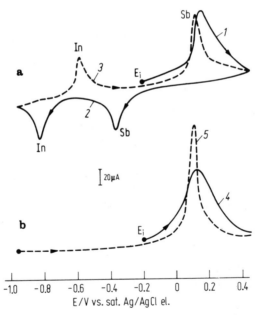

Figure 1.8. Anodic–cathodic–anodic (*a*) and anodic (*b*) potentiodynamic curves of the In–Sb alloy introduced into CPEE (*1, 2, 4*) and deposited on the electrode surface (*3, 5*). 2 M HCl; $v = 80$ m V/s; $\tau_{el} = 3$ min; $E = -1.0$ V (*b*), 55% In (*a*).

separately (curve *2*) on the electrode at the cathodic stage (under potentiodynamic conditions). Subsequently, on anodic dissolution, two signals are observed whose maximum potentials are close to the dissolution potentials of individual metals (curve *3*). Hence, a concentrate comprising individual metals is produced under potentiodynamic conditions.

In the second instance (Fig. 1.8b), subsequent to oxidation of the intermetallic compound (curve *4*), the deposit was obtained at a constant potential equal to -1.0 V. Under these conditions anodic curve *5* exhibits a single maximum at a potential close to the potential of InSb dissolution from the paste electrode. So, an InSb compound whose dissolution potential is close to the antimony dissolution potential is formed during electrolysis at a constant and sufficiently negative potential. As seen in Fig. 1.8, potentials of the maxima of InSb introduced in CPEE and produced by electrolysis are rather close. The shape of the anodic curves of the concentrates obtained on a graphite electrode from solutions containing antimony and indium ions depends on the concentration ratio of ions of these metals. Thus, on strong interaction of the metals, voltammograms recorded after potentiodynamic and potentiostatic conditions differ considerably.

Cadmium and bismuth form an eutectic alloy system, with the components being practically insoluble in each other at room temperature. Anodic dissolution of eutectic alloys introduced in CPEE is described by curves with two current maxima (curve of the first type, Table 1.1). Potentials of the anodic current maxima of equilibrium alloys are close to dissolution potentials of pure metals. The value of the maximum anodic current depends on the metal content in the alloy. The electronic work function ϕ changes linearly from ϕ_{Cd} to ϕ_{Bi} [63].

Figure 1.9 displays anodic–cathodic–anodic curves of the Cd—Bi alloy introduced in CPEE. It is easily seen that no principal differences are observed in the anodic behavior of the powdered alloy introduced in CPEE (curve *1*) and deposited using electrolysis on the electrode surface (curve *3*). In the latter case, however, potentials of the current maxima are shifted somewhat toward negative values. Oxidation of the concentrate formed on a graphite electrode surface during electrolysis of a solution containing Cd^{2+} and Bi^{3+} at the potential of the limiting diffusion current of Cd^{2+} is described by a curve analogous to curve *3* in Fig. 1.9. Different shapes of the curves describing dissolution of the alloy and the deposit obtained electrochemically can be explained by different energy states of the metals in the matrix of an equilibrium eutectic alloy and in their nonequilibrium conglomerates obtained through electrodeposition during cyclic polarization of CPEE (see Section 1.1.3), as well as by different distribution of metals in CPEE and on the electrode surface. Thus, in the absence of a strong interaction between metals, anodic

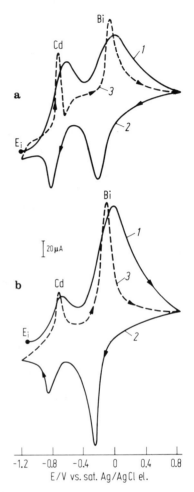

Figure 1.9. Anodic–cathodic–anodic curves of the Cd–Bi alloy introduced into CPEE (*1, 2*) and of oxidation of the deposit metals on the electrode surface (*3*) 0.5 M HCl; $v = 80\,mV/s$; 58.16% Cd (*a*), 36.8% Cd (*b*).

voltammograms recorded after the cathodic process under potentiodynamic and potentiostatic conditions differ little if at all.

So, the shape of cyclic anodic–cathodic and sesquicyclic anodic–cathodic–anodic curves of binary systems is determined by the nature of the system and conditions of its formation. Two independent peaks of the anodic current on the oxidation curves of binary alloys comprising weakly interacting metals make it possible to determine the ratio of the metals in the alloy. In the case of strong interactions, the first anodic curve furnishes information about the phase composition and the subsequent cathodic and anodic curves bear information about the elemental composition of the material, thus providing for the phase and elemental analyses of alloys.

Different shapes of the anodic curves of metals and alloys obtained by melt crystallization and of electrochemically formed deposits are indicative of different energy states of these entities, a circumstance which in itself may be a source of rather useful information.

One can clearly see the influence of the electrolysis conditions on the electrochemical deposit formation. Under potentiodynamic conditions the electrolysis leads to the formation of deposits consisting of individual metals, that is, separate deposition takes place. Electrolysis conducted at the limiting current potential of the electronegative component promotes interaction, for example, formation of IMC, this actually showing up in the subsequent anodic curve. The formation of metastable phases taking place under these condition (which is, incidentally, always the case with the SV method) leads to emergence of additional signals on the anodic voltammogram, a fact which makes it practically impossible to extract analytical information from the curve. It is necessary to take special steps aimed at decreasing the interaction.

The concepts developed herein provide for interpretation of experimental facts, allow one to choose a priori the analysis conditions, and testify to the principal possibility of developing methods of examination and analysis of solutions, solids, and alloys.

THE CHOICE AND FORMATION
OF ANALYTICAL SIGNALS

2.1. ANALYTICAL SIGNALS IN THE STRIPPING ANALYSIS
OF SOLUTIONS

The theory of stripping methods (SV, SCP, SCA) used to analyze solutions has been covered in sufficient detail in the monographs [15, 38, 29]. The first stage of the determination process is the stage of electrochemical or adsorptive accumulation (preconcentration) of the analyte on the surface or in the bulk of the electrode. Preconcentration is provided during a preset time at a given electrode potential, and hydrodynamic conditions that ensure a steady flow of the analyte to the electrode surface (rotating electrode, stirring of the solution, flow-through electrolyzers).

Brainina [38], Vydra and co-authors [15], and Wang [29] noted that the possibility exists of realizing various methods of the analyte preconcentration from the solution. Sorption from the gaseous phase can be viewed as a new method of SV concentration [64–67]. It is also possible to improve the detection limit of stripping procedures by using catalytic reactions [68].

Some problems concerned with discharge-ionization of metals on the surface of a foreign solid electrode are considered in Chapter 1 and in Section 2.2–2.5. Other types of concentration are discussed in more detail in Chapter 4.

In the SEAM, analytical information proper about the solution composition is obtained at the second stage of determination, where the deposited concentrate is oxidized or reduced. Depending on the regime of these processes, the analytical signal can be represented by:

- The maximum of the current of the concentrate electrochemical transformation with the potential scanned linearly (SV), including direct current (d.c. SV or LSSV), pulse (NPSV or DPSV), and alternating current (a.c. SV) modes of polarization
- The potential at a preset external current (SCP); in a particular case, the current may be equal to zero (PSA)
- The instantaneous current when $t \to 0$ at a preset potential (SCA)

In all of the techniques, the analytical signal can be the quantity of electricity, Q, equivalent to the amount of the concentrate that undergoes electrochemical transformation.

2.1.1. Stripping Voltammetry

Mathematical simulation of the processes of discharge-ionization of metals on a foreign surface show ([14, 38] and Chapter 1) that a functional dependence exists between the maximum current of oxidation of the metal accumulated on the electrode and the concentration of the metal ions in the solution. This dependence is direct for ionization of the microphase and is linear for ionization of the macrophase of the metal.

For mercury electrodes (HMDE, MFE, Hg/CE), activity of the metal is proportional to its concentration in the amalgam produced. As a result, a direct relationship is observed between the maximum of the registered current and the concentration of the analyte ions in the solution over a broad concentration interval.

The technique that has received the most study and that has been used most extensively for a long time is the linear scan of potential (d.c. SV or LSSV). This technique, which has been covered amply in the literature on the SEAM [14, 15, 38, 29], provides for sufficiently low detection limits ($\sim n \times 10^{-11}\,mol\,l^{-1}$) of ions of many metals and ensures an adequate resolution of responses ($\sim 100\,mV$).

Less known is a derivative variant of the LSSV, where the relation $di/dt = f(t)$ is registered, which allows one, at high potential scan rates ($> 100\,mV\,s^{-1}$), to improve the detection limit and to obviate the necessity in deaeration of solutions. The use of this SV variant will be considered in detail in Chapter 4.

To increase the signal–noise ratio compared to LSSV, small-amplitude pulses are superposed on the potential that varies linearly with time. Diversity of the pulse shapes and signal measurement techniques provides for a wide set of SV modifications. The most extensive use has been made of alternating-current (a.c. SV) and differential pulse (DPSV) voltammetry. Thanks to such measurement procedure, the contribution of the current capacity component to the registered current lowers, the detection limit is improved ($\sim n \times 10^{-11}$ $mol\,l^{-1}$), and the possibility appears of a rapid determination ($\tau = 1–3\,min$) of various elements at a level n $ng\,l^{-1}$ (Zn, Cd, Pb, In, Tl, Cu, As, Bi, Sb, Au, etc.).

The effect of surfactants on the kinetics of electrode processes shows up much weaker in DPSV than in a.c. SV, this allowing the time preconcentration to be increased ($\tau \geqslant 15–20\,min$). With the advent of carbon electrodes plated with mercury in situ, the DPSV method has received worldwide recognition. It is not by chance that every other work concerned with the use of SV in the analysis of natural and industrial materials exploits this SV variant. In the case of adsorptive concentration of organic substances at HMDE, more than 80% of the techniques employ DPSV (see Chapter 4).

The advantage of DPSV over other SV variants, specifically LSSV, becomes clear mainly when HMDE is used. With film electrodes and electrodes plated with mercury in situ, the advantage is not so significant. This is because the separation of the faradaic and capacity components in DPSV is based on the use of the different behavior of the time dependence of the diffusion (i_d is proportional to $1/\sqrt{t}$) and capacitive i_c is proportional to $1/t$) currents.

In the case of thin-film electrodes, the electroactive substance is localized practically in the surface layer. The time dependence of the substance electrochemical transformation current approaches the inverse-proportional one ($i \sim 1/t$) and the time selection turns out to be of low efficiency. Thus, Florence [69] showed that when the pulse amplitude was 25 mV, the detection limit of Cd, Pb, and Cu at GCE plated with mercury in situ is 3–5 times lower compared to LSSV. The authors [70, 69] hold the opinion that LSSV and DPSV are practically interchangeable and it is these methods that should be preferred in SV [71].

The theory of DPSV, the equipment used, and the experimental procedure are described in sufficient detail in the monographs [5, 16, 23, 72]. The applied aspect of DPSV is dealt with in numerous developments which are cited in Chapter 4.

2.1.2. Stripping Chronopotentiometry—Potentiometric Stripping Analysis

In the galvanostatic variant of stripping chronopotentiometry (SCP), the electrode with deposited concentrate is kept at a preset external current and the E–t curve is registered, on which a plateau of the potential that varies weakly with time corresponds to dissolution of the concentrate. In SCP, the analytical signal is the so-called transition time (τ_t) which is directly proportional to the concentration of metal ions in the solution over a broad concentration interval. Note that sensitivity of the method ($d\tau_t/dc^0$) is independent of the electrode process kinetics and, for solid-phase concentrates, of the state of the concentrate components on the electrode, a feature which is of great utility, especially in automated analysis.

Stripping chronopotentiometry has significant advantages over the direct variant of the method, where τ_t of successive electrode processes depends on τ_t of the preceding reactions, that is, on the concentration of electropositive components in the solution. In SCP, the preceding process is complete by the moment the potential of subsequent process sets up.

A SCP variant has gained in popularity, where dissolution of the concentrate proceeds without application of the external current and takes place because of the redox reaction between the concentrate and the solution component. The currentless variant of SCP was described in the 1960s (see, e.g., [73]). In the 1970s, Jagner and Granelli proposed the term PSA [74] for this variant.

In PSA, subsequent to the accumulation stage, the working electrode is disconnected from the power source and the dependence of the potential on the oxidation time (curve E-t) of the concentrate is registered. Oxidation reactions follow the sequence corresponding to the equilibrium potentials of the concentrate components, making it possible to estimate qualitatively and qnantitatively the composition of the concentrate and, hence, of the test solution, similar to SCP under galvanostatic conditions.

The PSA variant considered above is extremely convenient when a GCE or another substrate plated with mercury in situ is used. The presence of Hg^{2+} ions added in excess to the test solution ensures the formation of the mercury surface at the stage of preelectrolysis and oxidation of the amalgam at the stage of recording E-t curves.

The metal oxidation reaction rate is determined solely by the speed at which the oxidant is transported to the electrode surface. In this connection, of great importance in PSA are hydrodynamic conditions not only at the stage of preelectrolysis (as is the case in other SEAM), but also at the stage of the concentrate oxidation. The author [75] analyzed the effect of the oxidant flow velocity upon the analytical signal in PSA and proposed to distinguish between two cases:

- Normal PSA, where hydrodynamic conditions are the same at both stages of determination and the oxidation rate is controlled by the speed of solution stirring.
- Stationary PSA, where hydrodynamic conditions are different for the two stages. In this instance, the concentrate is dissolved without stirring of the solution (rotating of the electrode), that is, the oxidation rate is determined by the rate of the oxidant diffusion to the electrode surface.

By varying the rate of Hg^{2+} diffusion to the electrode surface at the expense of, for example, enhancing viscosity of the solution through introduction of organic additions (polyelectrolytes), the author [75] achieved a 50-fold improvement in the detection limit of Cu, Cd, Zn, Pb, Bi, Tl, and Ga determined by the PSA method; that is, it is possible to detect 0.01–0.10 ng ml^{-1} ions of the given metals with $\tau_d = 3$–5 min.

The limitation of the stationary PSA is the dependence of the results on the solution vibration, external mechanical interferences, and other factors affecting the oxidant flows.

Potentiometric stripping analysis was further developed in flow-through systems that employ thin-layer electrochemical cells [76].

The usual and flow-through variants of PSA are used currently for analysis of various natural waters. A detection limit of $0.3-0.8\,\mathrm{nmol\,l^{-1}}$ was attained in the determination of Cd, Pb, and Cu by the flow-through method.

Applications of the PSA method are given in Chapter 4.

2.1.3. Stripping Chronoamperometry (SCA)

Under conditions of solid-electrode SCA the instantaneous current ($i_{t\to0}$) of the metal microphase dissolution is proportional to the amount of the deposited metal or to the concentration of the metal ions in the test solution. For dissolution of the macrophase ($\gamma q_{II} \gg 1$), $i_{t\to0}$ is determined only by the kinetic parameters of the process and is independent of the concentration of electroactive ions in the solution. Thus, information about the concentration of ions in the solution can be obtained only in particular instances of the metal microphase dissolution. The absence of the $i_{t\to0} = f(c^0)$ relation in other cases makes SCA inapplicable for analytical purposes. Electrolysis at a preset potential can be probably used only in the coulometric variant which is adaptable also to all of the SEAM considered above.

At the present stage of development of the SEAM it is impossible to account quantitatively for all factors that influence the generation of the analytical signal in the analysis of solutions. A problem arises in providing for such conditions at the stages of the analyte accumulation and subsequent electrochemical transformation of the concentrate produced, where the effect of the interfering factors on the recorded analytical signal becomes statistically insignificant.

2.2. THE INFLUENCE OF THE ADATOM-INITIATED ELECTRODEPOSITION ON THE ANALYTICAL SIGNAL

2.2.1. Formation of a Mercury-Plated Graphite Electrode

One of the most effective methods used to suppress such processes as deposition of the analyte component in two energy states and the mutual effect of the components in the concentrate produced on the electrode, providing simultaneously for a lower detection limit in the SEAM, is the use of a mercury-plated graphite (or glassy carbon) electrode (Hg/CE) [77, 38] (see Section 3.3). However, formation of the Hg/CE in situ encounters some difficulties associated with the slow kinetics of mercury deposition on the graphite surface, a circumstance which leads to a delay in the formation of the mercury surface and can cause errors in the analysis of solutions.

Electrochemical deposition of mercury on graphite requires some overvoltage for nucleation to take place. The process is highly dependent on the surface condition [78] and electrolyte composition [14, 15, 78].

A mathematical model of growth and coalescence of mercury droplets on glassy carbon was proposed by Demin and co-workers [79] who, in addition, performed microscopic studies of the electrode surface. Changes of the surface, mainly of mercury, occurring during operation of a graphite electrode with mercury deposited in situ were studied in detail, including the microscopic level, by Zakharchuk and Valisheva [78]. They showed that during anodic polarization of the electrode, for example, in a chloride-containing medium, mercury does not dissolve but is covered with a film of a low-soluble compound which cannot be removed electrochemically. The authors [80, 81] came to analogous conclusions. However, these difficulties can be overcome to some extent if we limit the anodic polarization potential and do not bring it to the mercury oxidation potential.

More difficulties are involved with the initial stages of the electrochemical deposition of mercury on graphite. As was noted above, the deposition proceeds with a rather high overvoltage, this causing a delay in the coating of the electrode with mercury and a drift of the analytical signal (response) with time.

Taking into account the data given in Section 1.1, one should expect a change in the kinetics of the initial stages of mercury deposition in the presence of ions of metals with small electronic work function, which are capable of forming adatoms on the electrode surface. When a carbon electrode is used, this ability increases in the series (1.20) from the left to the right. Thallium can serve as the auxiliary element during electrochemical deposition of mercury on graphite, since thallium is the last element in the series (1.20) that tends to form adatoms. A characteristic voltammogram of thallium is shown in Fig. 1.1, where the peak at potential -0.8 V is due to the phase (Tl_{II}) oxidation and the current over the interval from 0 to -0.1 V results from oxidation of thallium adatoms (Tl_I).

Figure 2.1 presents signals of electrodissolution of thallium from the first (1) and second (2) energy states as a function of the electrode potential at the stage of deposition. As is seen from the figure, the signal and, consequently, the amount of metal M_1 attains a limiting value at potentials more negative than -0.85 V. The metal phase starts growing intensively at potentials more negative than -0.95 V. Figure 2.2 shows a dependence of the electropositive signal of thallium on the electrolysis time and the concentration of thallium ions in the solution. The deposition time dependence of the signal is described by a curve with saturation. The region of saturation corresponds probably of filling of the active electrode surface with adatoms (M_1). Depending on the concentration, the signal grows initially and then drops, a fact which is

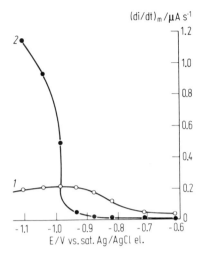

Figure 2.1. Electrode-potential dependence of the current time derivative maximum when thallium deposited on a graphite electrode is electrooxidized from the first (*1*) and second (*2*) energy states (0.1 M NaClO$_4$ + 0.3 M HClO$_4$ + 5 × 10^{-8} M Tl$^+$; $\tau = 5$ min; $v = 0.2$ V/s).

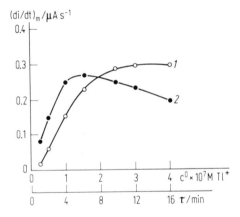

Figure 2.2. Maximum of the current time derivative corresponding to oxidation of thallium in the first energy state versus electrolysis time (*1*) and Tl$^+$ concentration in solution (*2*). 0.1 M NaClO$_4$ + 0.3 M HClO$_4$; $E = -0.85$ V; $v = 0.2$ V/s; $\tau_d = 5$ min (*2*); $c^0 = 1 \times 10^{-7}$ M Tl$^+$ (*1*).

probably due to appearance of the metal phase in the second energy state. It follows from Fig. 2.1 that if a solution containing less than 1×10^{-7} M monovalent thallium cation is electrolyzed for 5 min at a potential more positive than or equal to -0.85 V, the metal is deposited in the first energy state (adatoms).

Now consider deposition of mercury under these conditions [82]. Derivative anodic voltammograms (Fig. 2.3) were recorded subsequent to electrolysis of solutions containing Hg^{2+} and Tl$^+$ ions separately and together. It is readily seen that mercury is not deposited individually on the electrode surface from such a dilute solution. Electrolysis of the solution containing

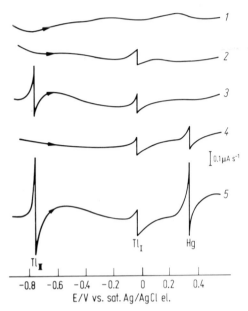

Figure 2.3. Derivative potentiodynamic curves of the ionization of deposits produced through electrolysis of solutions containing 0.1 M $NaClO_4$ + 0.3 M $HClO_4$ and 1×10^{-7} M Hg^{2+} ($1-3$), $5 + 10^{-8}$ M Tl^+ ($2, 3$), 5×10^{-8} M Hg^{2+} and 5×10^{-8} M Tl^+ ($4, 5$). $\tau_d = 5$ min; $v = 0.3$ V/s; $E_{el} = -0.9$ V (1); $E_{el} = -0.85$ V (s.c.e.) ($2, 4$); $E_{el} = -1.1$ V ($3, 5$).

mercury and thallium at a potential at which thallium is deposited only in the first energy state is accompanied by appearance of thallium and mercury anodic current peaks on the anodic voltammogram. The maximum oxidation current of mercury rises as the concentration of mercury ions in the solution is raised. The maximum oxidation current of thallium is independent of the mercury concentration. Thus, thallium initiates deposition of mercury on the graphite surface, while mercury has no influence on the discharge-ionization of thallium.

Photographs of the electrode surface in the solution with current applied and in air show that, in the presence of thallium adatoms, a more uniform distribution of mercury over the electrode is achieved, mercury droplets are smaller compared to the deposition of pure mercury, and no coalescence of mercury droplets is observed when the electrode is taken out of the solution (this testifying to a stronger attachment of mercury to the surface). Thus, a lower overvoltage is required for depositions of mercury on the graphite surface modified with thallium adatoms. Mercury is deposited from less-concentrated solutions, forming a finely dispersed coating which is more uniform and more stable than that obtained on a nonmodified electrode.

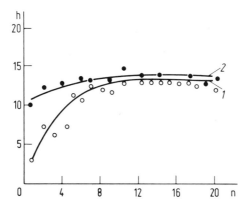

Figure 2.4. Response of lead as a function of the test solution composition: $1 - Pb^{2+}$; $2 - Pb^{2+}$, Cd^{2+}, Zn^{2+}, Bi^{3+}, Al^{3+} $(0.1$ M $HCl + 5 \times 10^{-5}$ M $Hg(NO_3)_2$; $E_{el} = -1.0$ V; $\tau_{el} = 1$ min; $c^0 = 0.5$ μg/l).

Initiation of mercury electrodeposition by thallium adatoms can probably be considered as a particular case of a more general phenomenon of adatom-catalyzed electrodeposition [82] or of the "substrate effect" [83].

The change in the conditions of the Hg/CE formation in the presence of ions of elements with a smaller electronic work function results in a lower time drift of the response and improved reproducibility of determinations.

It can be seen from Fig. 2.4 that when a graphite electrode plated with mercury in situ is used, the response of lead becomes constant (within the measurement error) much faster in the presence of auxiliary elements (AE) (which are represented by metal microimpurities) in a sample of natural water than in pure solutions of lead in HCl. As the number of successive stripping measurements made under the same conditions is increased, the amount of deposited mercury rises, this facilitating to a great extent the setting-up of a stable response of lead.

2.2.2.   Lowering the Detection Limit of Elements

Initiation of electrodeposition of elements featuring a larger electronic work function by adatoms of metals having a smaller electronic work function is also used to lower the detection limits of those elements with a large electronic work function. In the absence of auxiliary elements forming adatoms on the electrode surface, the maximum current versus electrolysis time curves show a delay in deposition of some elements on the graphite (or glassy carbon) electrode (with the delay decreasing as the concentration is raised), or no deposition is observed. If auxiliary elements are added, the character of the

Table 2.1. Detection Limits of Elements by the SV Method with the Use of Auxiliary Elements (AE)

Analyte	AE	c_{min} (M) w/o AE	c_{min} (M) with AE	Ref.
Pt	Pb, Cd	No signal	2×10^{-7}	84
	Hg		3×10^{-8}	85
	Sn		5×10^{-9}	86
	Pb		1×10^{-8}	87
In	Hg	No signal	6×10^{-10}	88
Rh	Hg	No signal	1×10^{-7}	89
As	Cu	1×10^{-5}	4×10^{-8}	90
	Au		8×10^{-8}	91
Te	Cd		1×10^{-8}	92
As	Cu		1×10^{-6}	93
	Hg		5×10^{-6}	94
Se	Bi	1×10^{-4}	1×10^{-8}	95
	Au		2×10^{-9}	96
	Cu		5×10^{-8}	97
	Pb, Cu		3×10^{-6}	98
	Cu	5×10^{-5}	3×10^{-8}	99
Te	Cu	1×10^{-7}	8×10^{-8}	99
Hg	Cu		2×10^{-9}	100
	Cd		2×10^{-9}	100
	Cd		2×10^{-9}	101
	Cu	4×10^{-7}	3×10^{-9}	102
	Pb		1×10^{-8}	102
	Cu		1×10^{-9}	102

principal dependences changes and, as a rule, a signal appears, which is related by the functional dependence to the content of the electropositive element (the one having a larger electronic work function), and the detection limit is lowered appreciably. The corresponding data are summarized in Table 2.1.

Comparing the data of the table and the series (1.20), we can see that the efficiency of the auxiliary elements changes, in general, symbatically with decreasing electronic work functions of the corresponding metals. Deviations are most likely due to different experimental conditions.

Thus, the role of auxiliary elements or of the substrate is played by the elements that stand in the series (1.20) to the right of the elements which are not deposited by themselves. The "substrate effect" enhances in most cases as the difference between the electronic work functions of the corresponding elements increases. Taking into account the data given above, electrolysis of the solution containing M and AE ions at the graphite electrode should be represented as follows. At the initial stage, the AE in the first energy state (adatoms) is deposited on the graphite surface, filling the entire or part of the surface active to the given AE depending on the electrolysis time, concentration of the AE ions in the solution, and the electrode potential. On the surface modified in this way, adatoms serve, most likely, as some crystallization centers or active sites, where deposition of elements proceeds much faster, probably owing to the formation of transition complexes.

Consequently, the interpretation of the initial stages of electrodeposition of metals and products of their interaction on the surface of a foreign electrode should take into account the following two aspects, namely, the thermodynamic aspect that manifests itself in the shift of the discharge-ionization potentials as a result of the metal–metal interaction, and the kinetic (catalytic) aspect that consists of acceleration of the process due to modification of the surface with adatoms. Kinetic effects show up at trace concentrations of the corresponding ions in the solution, making them often "uncontrollable" factors and leading to irreproducibility of the results both of analytical and electrochemical studies. However, the effect of adatoms, if taking place following a certain program, lowers the detection limit of some elements, and improves reproducibility. Initiation of electrodeposition by adatoms also shortens the time of analysis thanks to the elimination of delays in the Hg/CE formation and stabilization of finely dispersed mercury coating on the Hg/CE surface.

2.3. THE "THIRD"-ELEMENT EFFECT

Interaction of metals in binary and multicomponent systems (concentrates) that are usually formed under actual analysis conditions causes distortion of signals of individual elements.

The method which is widely exploited in SV to eliminate such interactions, that is, the use of Hg/CE, fails to be sufficiently efficient at all times. In these circumstances, it may be a good practice to add ions of a "third" elements to the solution to bind the interfering element in the concentrate. In this case a strong interaction between the analyte and interfering elements is replaced by a still stronger interaction between the interfering element and the "third" element. By choosing the correct concentration of ions of the "third" element and the electrolysis conditions, one can obtain an undistorted response of the analyte element.

Table 2.2 specifies "third" elements for some binary systems. It is shown [14] that as regards the degree to which the interference of, for example, copper and silver with responses of more electronegative elements is eliminated, the "third" elements can be arranged in the series $Ga > Fe > Sn > Zn \approx In >$

Table 2.2. Recommended "Third" Elements for Some Binary Systems in the SEAM [105]

Analyte	Binary System	"Third" Element	Electrode
Zn	Zn–Cu	Ga, Ge	Hg/CE
	Zn–Ag	Ga	Hg/CE
Cd	Cd–Cu	Sn, Hg	GE
	Cd–Sb	Cu	Hg/CE
In	In–Cu	Te	Hg/CE
Sb	Sb–Se	Hg	GE
	Sb–Te	Hg	GE
Se	Se–Sb	Cu	GE
Pb	Pb–Cu	Hg	GE
Te	Te–Cu	Hg	GE

Cd > Pb and Ga > Zn > Cd > Cu. These series correlate with the variation of the Gibbs energy when alloys are formed in the interfering element–"third" element systems.

In [103], the "third"-element effect is used to enhance resolution when additive superposition of responses in the SEAM is observed.

In some cases, for example, for the Sb–Se system [104], two different "third" elements can be used: mercury that eliminates the influence of selenium on the antimony response, and copper that allows determination of selenium at 50-fold amounts of antimony. The use of the described variant of the "third" element in combination with a properly chosen supporting electrolyte can improve appreciably the resolution of responses in the SEAM and optimize the analysis conditions. It should be remembered that the "third" element should not interact with the analyte element. Such a situation takes place, for example, in the Cd–Cu–Te system [105]. Copper impairs determination of cadmium, forming an intermetallic compound with the latter. In the presence of Te^{IV} a competing process consisting of the formation of a stronger copper–tellurium intermetallic compound takes place on the electrode. When the ratio Cu^{II}: Te^{IV} = 2:1, an undistorted cadmium response appears. If the Te^{IV} concentration is raised, formation of mixed copper and cadmium telluride probably starts on the electrode. Determination becomes impossible.

2.4. THE INTERNAL STANDARD METHOD

Results of measurements made in stripping methods of electroanalytical chemistry depend on a number of factors: the nature of the electrode; reproducibility and rate of formation of the working surface; the influence of the accompanying elements on the formation process; stability of the hydrodynamic conditions of the analyte elements ion transport to the electrode surface; the influence of the surfactants present in the solution and of low-soluble compounds which are liable to appear during concentration, measurement, and electrode regeneration [78, 106]. The diversity of the factors, including those that are usually unpredictable, leads to the situation where reproducibility (relative standard deviation) in the determination of the $10^{-8}-10^{-9}$ M concentrations as a rule exceeds 0.2–0.3 when the number of parallel measurements is not less than three. Besides, the drift of the response occurring on repeated measurements and caused by a slow formation of the mercury-plated electrode and nonuniform distribution of mercury over the electrode surface may be a source of systematic errors. Elimination of these errors makes the analysis substantially longer, since in this case a rather long time is required for the preliminary formation of the electrode surface.

The analysis algorithm used in the SEAM is as follows: (1) preparation of necessary reagents; (2) electrolysis of the solution being stirred; (3) rest time; (4) measurement of response (blank test); (5) regeneration of the electrode surface; (6) preparation of the test sample; (7) stages 2–5 repeated for the sample; (8) introduction of an addition in the test sample; (9) stages 2–5 repeated for the sample with the addition; (10) introduction of another addition in the sample containing the first addition; (11) stages 2–5 repeated; (12) calculation of the concentration sought.

As a rule, the mean value of the response taken over three measurements is used, that is, the measurement cycle (stages 2–5) is repeated three times. Thus, altogether 12 measurement cycles are carried out to analyze a single sample. The use of a reference element (internal standard) makes it possible to reduce the whole analytical procedure so that stages 2–7 have to be carried out only once. The reference element is chosen on the basis of the series (1.20) and the data of Table 1.1. The reference element should meet the following requirements:

- It should not interact with the analyte element in the electrochemical concentrate, that is, the difference between the electronic work functions of these elements should be small, with the relation $\phi_x > \phi_{ref}(\phi_x$ and ϕ_{ref} are the electronic work functions of the analyte element and the reference element, respectively) being preferable.

- The concentration of the reference element in the test sample should be known or be considerably less than the concentration of the analyte element. In the latter case the reference element is added at stage 1.

- Responses of the analyte and reference elements should be well resolved.

If a direct relationship is observed between the concentration and the

Table 2.3. Responses of Lead and Cadmium at a Glassy-Carbon Electrode with Mercury Plated Preliminary and in Situ (1×10^{-7} M Cd^{2+} and Pb^{2+} in 2 M HCl) [5]

Coating Plated	n	Mean Coating Thickness $\times 10^6$ (cm)	I, (μA) Pb	Cd	I_{pb}/I_{Cd}
Preliminary	1	1.3	18.7	13.7	1.36
	2	1.3	18.3	11.5	1.59
	3	1.3	17.3	10.3	1.67
	4	1.3	21.3	10.3	2.06
	1	11.0	22.0	20.0	1.10
	2	11.0	21.0	20.0	1.05
	3	11.0	20.5	20.3	1.00
	4	11.0	21.5	20.8	1.03
In situ	1	1.3	20.3	16.7	1.21
	2	2.6	20.0	18.5	1.08
	3	3.9	20.7	20.3	1.01
	4	5.2	19.6	19.7	1.00
	1	11.0	21.5	20.7	1.03
	2	22.0	21.5	20.7	1.03
	3	33.0	21.5	20.9	1.02
	4	44.0	21.5	20.7	1.03

Table 2.4. Results of Lead Determination Using a Reference Element with the Electrode Surface Regenerated by Different Methods ($c_{Pb^{2+},} = 1.8 \times 10^{-8}$ M)[a]

Electrochemical Regeneration						Mechanical Regeneration					
R_{Pb}	s_r	R_{Cd}	s_r	R_{Pb}/R_{Cd}	s_r	R_{Pb}	s_r	R_{Cd}	s_r	R_{Pb}/R_{Cd}	s_r
40	0.29	20	0.26	2.00	0.05	38	0.18	20	0.11	1.9	0.06
56		26		2.15		46		22		2.09	
56		27		2.07		44		22		2.0	

[a] R = response.

maximum electrooxidation current (appropriate conditions are chosen in the course of the technique development), the following simple relation between the concentrations of ions of the corresponding elements is realized:

$$c = Kc_{ref}I/I_{ref} \qquad (2.1)$$

where K is the proportionality factor; c_{ref} is the concentration of the reference element.

Let us consider by way of example the use of the internal standard method for the determination of the lead ion concentration with cadmium as the reference element (Table 2.3). It can be easily seen that Cd fulfills the requirements described above. The relation of the signals remains practically unchanged if determination conditions are altered. The data of Table 2.4 illustrate improvement in the reproducibility of measurements when a reference element is used. Whatever the method of the electrode surface regeneration, the relative standard deviation of the ratio of the signals is much less than that of the signals proper.

The use of the internal standard method makes it possible to curtail considerably the analysis time (3–5 min instead of 30–50 min) and to improve metrological characteristics of the results obtained. These peculiarities result because the standard signal (signal of the reference element) is shaped under the same conditions as the signal of the analyte element. All factors (including unpredictable ones) have practically the same effect on both signals.

The internal standard method was exploited for simultaneous determination of the concentrations of copper, lead, and cadmium ions in sea and river waters. Cadmium was used as the reference element [107].

One paper [108] describes application of the method for the determination of Cd, Pb, and Cu with indium as the reference element.

2.5. ANALYTICAL SIGNALS IN THE STRIPPING ANALYSIS OF SOLIDS

The variants of the SEAM considered in this section differ from the previously described ones in that they do not necessarily call for the preliminary concentration stage. Thus, in the phase and elemental analysis the source of information is the current of the electrochemical transformation of the phase which is already present in the substance studied. When studying specific features, particularly defects in the crystalline structure of a solid, the source of information is represented by the currents of electrochemical transformation of oxygen and hydrogen adsorbed on the surface of the substance investigated. In some cases, adsorbates are formed through preliminary electrolysis at the water decomposition potential. These processes will be considered in detail in Chapter 6. Here we consider only some aspects involved in generation of a response with the use of a carbon-paste electroactive electrode (CPEE).

Potentiodynamic conditions are most generally used for polarization of CPEE. In this case a clear-cut dependence of current on potential is observed, which can be easily measured (SV variant). When potentials of electrochemical transformations of the investigated substances are sufficiently different and the processes themselves are electrochemically reversible, galvanostatic conditions of electrode polarization (stripping chronopotentiometry) can be used, making it possible to simplify measurement devices. The analysis and the study of systems comprised of components with close electrochemical properties often require that signals are recorded under potentiostatic conditions (stripping chronoamperometry), because it is impossible to choose the initial potential for recording the potentiodynamic curve such that no changes are incurred in the system studied.

Thus, the choice of the polarization conditions is determined by the nature of the system investigated.

Carbon-paste electroactive electrodes are usually used to study solids. A CPEE is a system comprised of an electroactive low-soluble substance being studied, which is introduced into the paste of a carbon powder, and a binding liquid [109]. The CPEE operates within the range of potentials of a carbon paste electrode [14, 21, 109, 110]. The paste serves as the conducting medium with uniformly distributed particles of the substance capable of electrochemical transformations. The electrode process proper is localized at the electrode-solution interface if a nonconducting organic liquid is used as the binder, or covers the electrode volume when the electrolyte solution serves as the binding liquid.

Four types of electrode reactions at the CPEE are described: (1) oxidation and reduction of a solid accompanied by formation of soluble products which diffuse into the bulk of the solution or enter into a subsequent chemical reaction whose products are also soluble; (2) oxidation or reduction of solids accompanied by formation of another solid; (3) reduction of oxygen adsorbed from the gaseous phase or from the solution; (4) electrochemical transformations of the electrolyte oxidation or reduction products adsorbed on the surface of the solid involved. Table 2.5 presents equations relating experimental and kinetic parameters of the process of the first type.

Electrochemical reactions of other types have no mathematical simulation, a fact which is due to difficulties encountered in finding an adequate model describing transformations of various classes of substances in the solid phase.

Table 2.5. Calculated Calibration Characteristics of Metal Oxidation from CPEE [14]

Conditions	Relationship Registered	Process	Equations Relating Basic Parameters
$E = E_i + vt$	$i = f(E)$	Reversible	$I = 0.29\ [(nF)^2/RT] \times vSdc_0$
		Irreversible	$I = 0.37\ [(nF)^2/RT] \times vSdc_0$
$i = \text{const}$	$E = f(t)$	Reversible and irreversible	$\tau = nFSdc_0/i$
$E = \text{const}$	$i = f(t)$	Irreversible	$i_{t-0} = nFSk_s c_0$ $\times \exp[\beta nF(E - E^0)/RT]$

Only results of experimental studies are known [14, 21, 110], which show that an analytical signal can be registered and this depends on the degree of dispersion of the substance analyzed and its concentration in the paste (I, τ_t, or $i_{t\to0}$). In practice, the maximum current which is observed under potentiodynamic conditions of the CPEE polarization is most frequently used as the analytical signal.

ELECTRODES AND ELECTROLYZERS

Development of stripping electroanalytical chemistry is connected with the use of metal electrodes, mainly mercury (hanging mercury-drop and mercury-film) electrodes, and carbon-containing electrodes. Mercury electrodes are amply described in the literature, in particular in monograph [15]. In recent years, HMDE have been subject to changes only at the design level. As far as solid electrodes are concerned, their perfection consists of the development of microelectrodes, searching for and using new electrode materials, improving the surface condition and methods of the surface regeneration, searching for new reactions, for example, over the interval of rather positive potentials, which cannot be realized with mercury electrodes [111–114]. Studies performed in this field before 1985 are covered in a monograph [28].

Let us focus on some studies made over the last five years. The requirements imposed on electrodes are as follows: electrochemical inertness over a broad interval of potentials; high overvoltage of hydrogen and oxygen evolution; low residual current (absence of pores and pronounced roughness of the surface); low ohmic resistance; the possibility of a sufficiently simple regeneration of the surface. All these factors should provide for high accuracy, sensitivity, and reproducibility of results and a low detection limit.

As is known, the response is due to the processes taking place on the electrode surface, that is, the processes depend on the surface condition. The surface depends on the nature of the electrode material and on the method of the electrode treatment (mechanical, chemical, or electromechanical). Specific chemical or electrochemical treatment can yield a modified electrode selective to ions and compounds of a certain type and representing a rather useful tool in electroanalysis.

3.1. METAL ELECTRODES

Metal electrodes considered in this section are also referred to as "indifferent" electrodes, but we shall not use this term since in actual fact the electrode surface seldom remains indifferent to the process proceeding on it. The electrode interacts, for example, with deposited metals which form adatoms on the surface (see Chapters 1 and 2). Formation of adsorption and phase layers alters the nature of the surface. In some processes, for example, during concentration of anions, the electrode material enters the concentrate composition.

Mercury Electrodes. The advantages of mercury electrodes over electrodes made of other materials are high hydrogen overvoltage and good reproducibility of the surface; failings of mercury electrodes are toxicity of mercury and low redox potential, which limit drastically the field of their application. Mercury electrodes are used principally for the determination of amalgam-forming metals.

Let us emphasize two modifications of mercury electrodes, namely, stationary hanging mercury drop electrodes and film electrodes.

The main cause of irreproducibility of the determination results obtained on stationary mercury hanging drop electrodes is the ingress of electrolytes to the capillaries. This undesired effect can be eliminated if one uses teflon or glass capillaries covered with polyethylene or silicon film [17]. There are electrode designs providing for a forced feed of mercury, the so-called "static mercury elecrode" which can operate either as a dropping electrode or as a stationary electrode.

In Ref. 115, a description is given of an electrode in the form of a "sessile" mercury droplet with dropping period of 2–6 min.

At present, some companies, and Metrohm in particular offer good constructions of stationary mercury electrodes.

Electrodes Made of Noble Metals. The use of these electrodes in the SEAM is limited because of a rather low hydrogen overvoltage at platinum, gold, rhodium, and palladium electrodes or at electrodes made of alloys of these metals. Another limitation of the given electrodes is the formation of oxide layers or dissolution of metals, for example, of gold, at potentials more positive than $(+0.9)–(+1.0)$ V (sat. Ag/AgCl electrode). The way such layers influence electrode processes is often hard to predict.

A complicating factor is also the interaction of the electrode metal with the analyte metals deposited during concentration, which can give rise to a systematic error of determination.

Out of the metals enumerated, platinum is used most frequently as the electrode material. Application of platinum is limited by a region of positive potentials and therefore such electrodes are supplementary to mercury electrodes.

Gold disk electrodes are used, in particular, for determination of tellurium-(IV) (detection limit $1–4 \times 10^{-9}$ M) [116], arsenic(III) (4×10^{-10} M) [117] and cadmium(II) [118]. A gold cylinder electrode was used to determine trace copper [119].

Electrodes Made of Other Metals. Silver electrodes are used as a substrate for a mercury film electrode or as an electroactive electrode whose material forms low-soluble compounds with analyte anions, for example, sulphide-ions [38]. Silver electrodes serve rather rarely as indifferent electrodes, specifically, for concentration and determination of Fe, Co, and Ni [120]

Discharge-ionization of lead on semiconductor n- and p-germanium electrodes was studied [42]. It was proposed to operate these electrodes in stripping voltammetry for the determination of the content of lead and cadmium ions. A study was also performed on the electrochemical behavior of lead at a liquid gallium electrode [121].

It was shown that the rotating disk lead electrode can be used in electroanalytical chemistry [122]. In 1 M H_2SO_4 the electrode features the working interval of potentials from -0.57 to -1.2 V (sat. cal. el.) and can be used to determine cadmium and thallium (detection limit 2×10^{-6} M).

A sensor with solid renewable electrodes made of such metals as Zn, Cd, Pb, Cu, Sn, Bi, In, Al, Ti, Ni, As, Au, and Pt was proposed for automatic electroanalysis of solutions. The results for the copper electrode [123] were the best.

3.2. CARBON ELECTRODES

Electrodes that enjoy the most use in stripping electrochemical methods are those made of carbon materials, such as glassy carbon, pyrographite, impregnated graphite, carbon fiber, and, in some cases, carbon glass ceramic [4, p. 289; 125, 124].

Successful application of carbon electrodes in the SEAM result from the high chemical and electrochemical stability of carbon materials, a relatively high hydrogen and oxygen overvoltage on the materials, a broad working potential range, simplicity of mechanical renewal of the electrode surface, and availability of carbon materials. Carbon electrodes do not interact with deposited metals during electrolysis, a feature which rules out the appearance of a systematic error caused by such interactions.

Glassy carbon [4, p. 289] or impregnated graphite [14, 38] is generally used. Glassy carbon is practically gas-tight, exhibits an extremely low porosity, conducts, electricity, and contains trace amounts of gaseous and other impurities. Glassy carbon possesses isotropic properties and does not require a particular orientation in the electrode device. It is necessary to strictly observe carbonization conditions in the course of glassy carbon production [4, p. 289].

There are problems connected with irreproducibility of the physicochemical properties of glassy carbon produced by different methods (different volume structure and degree of compactness affecting both bulk and surface properties).

Pyrolytic graphite is less common and represents a hard poreless, chemically stable, liquid- and gas-tight material with clearly pronounced anisotropic properties. Anisotropy makes it essential that a proper orientation of pyrographite be strictly observed when this material is used as an electrode [4, p. 289]. The surface parallel to the graphite layers should face the solution, a condition which is difficult to realize during electrode manufacture.

Carbon glass ceramic [125] is very close in its characteristics to glassy carbon. This material features low resistivity ($n \times 10^{-5}$ Ωm) and is practically gas-tight.

The glassy carbon, pyrographite, and carbon–glass–ceramic electrodes differ little in thier electrochemical properties and fields of application. These electrodes should be considered most probably as interchangeable.

The most extensive use has been made of impregnated electrodes comprising spectrally pure carbon impregnated with wax [4, p. 289], a mixture of paraffin and polyethylene [14], paraffin and polystyrene [4, p. 289], and epoxy resins [14, 124, 126]. A noticeable decrease in residual currents is observed when impregnated graphite electrodes (IGE) are used [14]. Impregnated graphite electrodes are machined more readily than glassy-carbon and pyrographite electrodes, because graphite is softer than glassy carbon.

Organic reagents adsorb better on graphite surfaces than on glassy carbon, a fact which makes it possible to use the electrode for adsorptive concentration reactions. Besides, there is evidence that the active sites for transition of electrons on glassy carbon are graphite edges inherent in the GC structure [127, 128].

Carbon paste electrodes (CPE) are more or less extensively used both for analysis of solutions [38, 14; 4, p. 289] and, probably, on a wider scale for the investigation and phase analysis of solid compounds [14, 21]. Simplicity of preparation and renewal of the electrode surface is by far the leading virtue of the CPE.

A very interesting and useful characterization of the microdistribution of conductive and insulating regions of CPE using scanning tunneling microscopy was presented by Wang and co-authors [129].

Use is made of electrodes produced by pressing or setting of a mixture of a binder and a graphite filler [130]. Various silicon, epoxy, polysulphide, polyethylene, and polystyrene resins and polymers serve as binders. Low residual currents are observed at such electrodes; the electrodes can be used over a wide range of potentials. The electrode surface lends itself to an easy renewal by cutting.

In recent years, carbon fiber [131, 132] has found an ever-increasing application for analytical purposes. Electrodes are made of a bunch of parallel graphite fibers. The bunch of fibers is encapsulated in an insulating matrix (polystyrene or epoxy resin) so that the fibers are insulated from each other. The electrode working surface is formed by cutting the rod across the fibers. Carbon fibers oriented in this way ensure a uniform distribution of the potential over the entire working surface. Residual current at a carbon-fiber electrode (CFE) is much lower than at glassy carbon, a circumstance which provides for a lower detection limit. Besides, the CFE is simple in manufacture, requires little time, and is relatively inexpensive [131]. A comparison of the reproducibility, sensitivity, and working potential intervals of the CFE of different types shows that CFE made of single fibers about 1–10 μm thick (microelectrodes) possess the highest electrochemical properties [131].

3.3. FILM ELECTRODES

Film electrodes include electrodes produced by deposition of some material on an inert electroconductive substrate (metal, carbon material). Film coatings are deposited chemically or electrochemically [133, 134] or the film is sputtered in a vacuum [135]. Depending on the material of the substrate and that of the film coating, the latter forms a more or less uniform thin layer. Mercury is deposited as a uniform film only on amalgam-forming metals and therefore a silver [133] or gold [134, 135] film is often deposited first on a glassy carbon [133], carbon fiber [135], or steel [134] substrate. Silver [136] and gold [137] are also used as a substrate in mercury film electrodes.

The disadvantages of mercury film electrodes on metal substrate are unstable thickness and composition of the mercury film as a result of mercury penetration into the bulk of the metal and formation of different concentration amalgams and possible interaction of the analyte metals deposited on the electrode with the substrate metal.

Preference is therefore given to carbon electrodes with mercury deposited preliminary or in situ [14]. In Russian literature such electrodes are known as mercury-plated graphite electrodes (Hg/GE). As already noted, mercury hardly ever forms a uniform film on carbon electrodes. Mercury(II) ions discharge primarily on the surface microdefects (microscratches, spallings, and cracks). As is seen from microphotographs, mercury microdroplets are clearly grouped near surface defects. The droplet dimensions depend on the electrode potential and decrease as the potential deviates from the mercury zero-charge potential.

It was shown that on platinum microdisks whose radius does not exceed 2.5 μm, mercury is deposited as a film on which droplets of hemispherical and spherical shape grow [138].

The Hg/GE surface generated in situ differs from the one produced preliminary by a more uniform distribution of mercury droplets, which is probably due to the initiation of the mercury deposition by adatoms of metals that discharge simultaneously with mercury (see Chapter 2).

Impregnated [4, 14, 126, 139] and pressed [130] graphite electrodes, glassy carbon [133, 140–144], and carbon fibers [145, 146, 147] are used as the substrate for Hg/GE plated with mercury in situ. A graphite substrate comprising 65 independent microdisks arranged as a circle and representing end faces of graphite fibers with a diameter of 9 μm is described in [147]. It is found that anodic current peaks are pronouncedly better at that electrode than at a normal flat electrode and are shifted toward more negative potentials.

The development of graphite electrodes plated with mercury in situ is one of the major advances in stripping analysis. A graphite electrode combines the advantages of solid and mercury electrodes: it operates over a broad working interval of potentials and its surface can be adequately reproduced. Intermetallic interactions do not, as a rule, show up at this electrode. Strong intermetallic interactions can be eliminated if one uses a graphite electrode with a preliminary deposit of mercury or introduces a "third" element into the solution. A graphite electrode plated with mercury in situ is less sensitive to surfactants than solid and mercury electrodes are.

A modification [148–150] of a carbon electrode is available which is coated with a film of a low-temperature anisotropic pyrolytic graphite; the film is deposited under controlled conditions on the surface of glassy carbon disks [148–150] or graphite rods [149]. Film graphite electrodes (FGE) can replace glassy carbon electrodes, which require long multiple and complex mechanical and electrochemical preparation for measurements. The FGE does not practically differ from glassy carbon when used as the substrate for mercury-plated graphite electrodes [148].

There are techniques of preparing a colloidal-graphite film on the end face of a metal rod (Al, Cu, Pt) [151, 152] or on a carbon paste electrode [153, 154]. In this case characteristics of the carbon paste electrode are considerably improved: residual current is lowered and reproducibility of results increases. An electrode with sputtered graphite is characterized by a high-energy homogeneity of the surface and operates successfully as a mercury-plated graphite electrode [151, 152, 154]. Similar characteristics are typical of electrodes obtained by application of a graphite suspension in an epoxy resin on the surface of a metal base [155] or plastic and glass rods [132] exploiting the technique of multilayer application and setting.

Platinum and tantalum electrodes coated with a film of electroconductive polypyrrole have been developed [156].

Electrodes coated with a semipermeable membrane for decreasing the influence of adsorption of organic surfactants on analysis results [144, 157] can be classified as a modification of film electrodes. The membrane is usually made of cellulose acetyl, ion-exchange resins, or cellophane [144, 158]. The response value is limited by the film permeability [144]. Recommendations on application of membraneous mercury film electrodes for analysis of natural waters with the organic component concentration varying over wide limits are given in [157].

3.4. MODIFIED ELECTRODES

In the last few years great interest has been aroused in electrodes with surfaces modified through chemical reaction, adsorption, or formation of a polymer film [159, 160]. For modification, the surface is treated with trialkoxysilane or trichlorosilane; the surface of carbon electrodes is chemically or electrochemically oxidized to carbonyl or carboxyl groups on which catalysts (e.g., cobalt or iron porphyrins) can be attached using substitution reactions [161, 162].

Electrodes whose surface is modified with electroconductive polymers, especially polypyrrole, have recently found extensive application [156, 163, 164, 165]. Polypyrrole is used also for modification of microelectrodes [166]. Heineman and co-workers were the first to employ an electrode coated with an electrochemically grown polypyrrole film in a flowing liquid to detect otherwise electroinactive anions [163, 167].

Another type of modified electrode is presented by carbon electrodes covered by or containing an ion-exchange polymer [168–170], an enzyme [171], or an extractive agent [172] reagent selective to analyte ions [173, 174, 175] or catalyzing an electrochemical reaction [173, 176–178]. A polymer, for example, chemically synthesized Cu(II)-containing poly(3-methylthiophene) powder, is sometimes pressed to form pellets which are used as detectors [179].

The authors [180] describe bioaccumulation and determination of the concentration of copper ions using a carbon paste electrode modified with algae [180].

A crown-ether modified carbon paste electrode has been used for preconcentration and determination of mercury [181], silver [182], and cadmium [183].

Some papers have been published, where the electrode surface is modified with immobilized enzymes [184] or a membrane is set up at the electrode–solution interface [157]. The use of modified electrodes opens up wide possibilities of improving analysis selectivity due to the reagent localized on the electrode surface.

Modified electrodes may be divided into several groups that differ in the mechanism of the analyte element concentration. The first (and the largest) group includes electrodes at which complex-formation reactions take place Table 3.1a,b). The second group covers electrodes at which concentration proceeds due to ion exchange (Table 3.1c). Extractive concentration is typical of electrodes belonging to the third group (Table 3.1d). In all the cases information about the analyte element concentration is furnished by oxidation or reduction currents of the complex, ionic associate or extract formed at the preconcentration stage. A large group of electrodes, for which the modifying agent is a catalyst of electrochemical reactions [227], serves mainly for the determination of organic substances. These electrodes are beyond the scope of this book, since they are used principally in direct (not stripping) voltammetry.

The papers [34, 33] are concerned with modified electrodes whose generation combines the stages of the reagent adsorption and the concentration of the analyte element on the electrode surface. It was proposed that this type of electrode be known as the "electrode modified in situ" (Table 3.2). The surface of the electrode can be modified with adsorbing organic substances [234], inorganic complex ions [34, 240–243], or metal adatoms [82].

The surface of graphite electrodes is modified with triphenylmethane dyes for determining the concentration of antimony and iodide ions (detection limit $10^{-8} - 5 \times 10^{-9}$ M) [234] or through anion-induced adsorption of mercury iodide complexes for the concentration and determination of $10^{-9} - 5 \times 10^{-10}$ M arsenic and tellurium [34, 240, 241]. Taking the thallium–mercury–graphite system as an example, it has been shown [82] that deposition of thallium adatoms initiates the mercury deposition process. Here the detection limit of mercury is lower and mercury is distributed more uniformly over the graphite surface, thus improving conditions of Hg/GE generation. Surface modification in situ is beneficial also in that it obviates the need for a preliminary formation of the surface (which is often a time-consuming operation), accelerating and simplifying the analysis.

At present, in situ modified electrodes have wide application in stripping electrochemical analysis (Table 3.2).

3.5. METHODS OF PREPARATION AND REGENERATION
OF THE ELECTRODE SURFACE

Among the numerous problems of electroanalytical chemistry is the problem of forming the electrode surface which possesses the required and reproducible properties.

Any preparation of the electrode surface starts with mechanical treatment.

Table 3.1. Electrodes Modified Preliminary

a. Modified with Complex-Forming Reagents

Electrode	Method of Modification	Analyte	SEAM Variant	Detection Limit (M)	Ref.
GCE	Formation of $\left[\begin{array}{c} O \\ O \\ O \end{array}\!\!-Si(CH_2)_3NHCOCOOH\right]$ surface-bound groups on glassy carbon electrode	Cu^{2+}	DPASV	1.5×10^{-8}	185
SMDE	Adsorption of pyrocatechol from buffer solution (pH 7.7) containing pyrocatechol	Cu^{2+}	CSV	1.5×10^{-10}	186
GCE	Adsorption of quinhydrone from solution containing $0.5\,M\,H_2SO_4 + 5 \times 10^{-3}\,M$ quinhydrone	Ascorbic acid	ASV	10^{-2}	187
GCE	Formation of Ru(III) and Ru(II) cyanide film from $2 \times 10^{-3}\,M\,RuCl_3 + 2 \times 10^{-3}\,M\,K_4Fe(CN)_6 + 0.5\,M\,NaCl$ (pH 2) using cyclic polarization over the interval 0.35–0.85 V (n.c.e.)	As^{3+}	ASV	5×10^{-6}	188
SMDE, GCE	Adsorption of trioctylphosphine oxide (TOPO) from $5 \times 10^{-3}\,M$ TOPO solution	U^{6+}	CSV	10^{-9}	189, 197
GE	Adsorption of malachite green (MG) from solution containing $0.5\,M\,KCl + 1.5 \times 10^{-5}\,M$ MG	Sb^{3+}	CSV	8×10^{-10}	190
Hg/CE	Adsorption of dithizone from phosphate buffer solution	Cu^{2+}, Cd^{2+}	CSV	10^{-7}	191
Pt	Adsorption of allylamine from 0.5% solution of allylamine	Ferrocen carboxaldehyde	ASV	10^{-7}	192
GPE	Adsorption of dizarine	Al^{3+}	AdPSV	1.5×10^{-7}	193
GCE	Coating of electrode with polypyrrole-N-carbodithionate using electropolymerization	Cu^{2+}	CSV	$n \times 10^{-6}$	194, 195

Table 3.1. (*Continued*)

Electrode	Method of Modification	Analyte	SEAM Variant	Detection Limit (M)	Ref.
CPE	Introduction of 2,2′-dithiodipiridine into the paste	Ag^+	CSV	8×10^{-9}	196
GCE	Adsorption of polyacrylamideoxime	Ag^+	CSV	1×10^{-8}	198
CPE	Introduction of dimethylglyoxime into CPE	Ni^{2+}	CSV	8×10^{-10}	199, 200
GE	Adsorption of poly(4-vinylpyridine) film containing bathophenanthroline acid	Fe^{2+}	CSV	1×10^{-8}	201
GCE	Electrochemical reducing polymerization with formation of film with imino-dicarbonic groups	Ag^+	ASV	10^{-7}	202
CPE	Introduction of dithizone into paste	Au^{3+}	DPCSV	5×10^{-8}	203
CPE	Introduction of decyl and dodecyl mercaptans into paste	Bi^{3+}	DPCSV	$n \times 10^{-8}$	204
GCE	Adsorption of trioctylphosphine oxide	U^{6+}	CSV	1×10^{-8}	205
CPE	Introduction of dry algae (0.5 g) into 1.5 g mineral oil + 2.5 g carbon	Cu^{2+}	CSV	2×10^{-6}	180
CPE	Introduction of dimethylglyoxime into paste	Ni^{2+}	CSV with a medium exchange flow system	1×10^{-8}	206
GCE	Adsorption of oxine	Tl^+	CSV	5×10^{-10}	207
GCE	Coating of electrode with Nafion film containing dicyclohexil-18-crown-6	Tl^+	ASV	2×10^{-12}	208
CPE	Introduction of dibenzo-18-crown-6 into vaseline oil —CPE	Hg^{2+}	Cyclic SV	2×10^{-6}	181
CPE	Introduction of hydrated zeolite into paste	Ag^+	CSV	7×10^{-7}	209, 210

Electrode	Procedure	Analyte	Method	Detection limit	Ref.
GCE	Application of 1 ml ethanol of tri-n-octylphosphine oxide onto electrode, drying using IR lamp, polymerization	Th^{4+}	CSV	1×10^{-9}	211
CPE	Introduction of dry bananas into paste	Dopamine	ASV	9×10^{-5}	212
CPE	Introduction of dry alga into paste	Au^{3+}	DPCSV, CSV, Cyclic SV	1×10^{-4}	213
GCE	Coating of electrode with Nafion film containing dicyclohexil-18-crown-6	Ag^+	ASV	1×10^{-8}	214
GCE	Adsorption of tri-n-octylphosphine oxide on electrode	Tc^{4+} Tc^{7+}	DPCSV	1.8×10^{-8}	215
CPE	Introduction of tri-crown ether into CPE by grinding with carbon in mortar	Ag^+	Cyclic SV	2.5×10^{-6}	182
Au/GCE	Application of 5 ml of solution containing $5 \cdot 10^{-3}$ M tri-n-octylphosphine oxide + 10% polyvinyl pyridine onto electrode surface, removal of solvent at 70°C using UV lamp	Hg^+	CSV	1×10^{-10}	175
CPE	Introduction of diphenylcarbazone into paste	Mn^{2+}	Derivative CSV	2×10^{-8}	216

b. Modified with Polymer Membrane

Electrode	Procedure	Analyte	Method	Detection limit	Ref.
GCE	Application of 5 µl of 5% methanol solution of cellulose acetate, evaporation of solvent, alkaline hydrolysis in 0.07 M KOH	Chlorpromazine	ASV	1.3×10^{-8}	217
Hg/CE	Application of 10 µl tributhyl-phosphate solution of tri-n-octylphosphine onto electrode surface, evaporation of solvent, polymerization at 70°C using IR lamp	Cu^{2+} Pb^{2+} Cd^{2+} Zn^{2+}	DPASV	$n \times 10^{-9}$	133

Table 3.1. (*Continued*)

Electrode	Method of Modification	Analyte	SEAM Variant	Detection Limit (M)	Ref.
Hg/CE	Adsorption of cellulose acetate from 5% solution, drying in air, alkaline hydrolysis in 0.07 M KOH	Cu^{2+} Pb^{2+} Cd^{2+} In^{3+}	ASV	1×10^{-9}	144
Hg/CE	Nafion polymer coating	Pb^{2+} Cd^{2+} Cu^{2+}	ASV	5×10^{-10}	218
Hg/CE	Coating of electrode with Nafion solution, followed by solvent removal	Bi^{3+}	ASV	1×10^{-11}	170
GCE	Coating of electrode with Nafion film containing 18-crown-6	Pb^{2+}	ASV	5×10^{-10}	183
	c. Modified with Ion-Exchange Polymers				
CPE	Introduction of Dowex-type cation-exchange resin into CPE	Cu^{2+}	CSV	2×10^{-6}	219
Pt	Adsorption of ion-exchange poly(4-vinylpyridine) polymer from 0.4% methanol solution	Cr^{6+}	CSV	10^{-8}	220
CPE	Introduction of liquid anion-exchange resin Amberlite LA2 into paste	Ir^{4+} $[IrCl_6]^{2-}$	DPCSV	5×10^{-7}	221

CPE	Introduction of poly(4-vinyl-pyridine) with counterions $[Fe(CN)_5L]^{3-}$, where L is pyridine-4-carboxyl	Primary amines	ASV	3×10^{-5}	222
CPE	Introduction of Dowex 50W-X8 ion-exchange resin in cation form into paste	Cu^{2+}	CSV	2×10^{-8}	223
GCE	Adsorption of polymers Eastman AQ-29D, AQ-55D from acetone solution	$[Ru(NH_3)_6]^{3+}$ $[Ru(bpy)_3]^{2+}$ MV^{2+} (methylviologen)	ASV	4×10^{-10}	168
CPE	Introduction of "damp and ground" ion-exchange resin into paste	Cd^{2+}	CSV	8×10^{-9}	169
CPE	Introduction of Amberlite A-2 ion-exchange resin into paste	Au^{3+}	CSV	5×10^{-6}	224

d. Modified with Extractants

CPE	Introduction of tri-n-octylphosphine oxide or tributhylphosphate into paste	Au^{3+}	DPCSV	1×10^{-7}	225
CPE	Introduction of chemical extractant N_{235}-alkyl tertiary amine N $(C_nH_{2n+1})_3$; n = 8-10 into paste	Au^{3+}	DPCSV	5×10^{-8}	172
CPE	Introduction of TTA extractant into paste	Tc^{4+}	ASV	4.5×10^{-9}	226

Table 3.2. Electrodes Modified in Situ

Electrode	Composition of Solution	Analyte	SEAM Variant	Detection Limit (M)	Ref.
HMDE	0.1 M NaClO$_4$ or 0.01 M acetate buffer solution and 1 × 10^{-4} M thiourea	Cu^{2+}		10^{-7}	228
Hg/CE	0.2 M NH$_4$Cl + 1% dimethylglyoxime solution	Co^{2+} Ni^{2+}	CSV	8 × 10^{-8} 5 × 10^{-7}	229
Au	(0.001 – 0.8) M H$_2$SO$_4$ + 5 × 10^{-2} gl^{-1} tetraethylammonium iodide	Cd^{2+}	Derivative ASV	2 × 10^{-8}	118
HMDE	0.1 M (NH$_4$Cl + NH$_4$OH) pH 8.3, 6 × 10^{-5} Malizarin red S	Ge^{4+}	CSV	1.4 × 10^{-9}	230
HMDE	Britton-Robinson buffer (pH 2), 1.8 × 10^{-4} M phenylene-diamine	Se^{4+}	d.c.SV	4 × 10^{-10}	231
CPE	10^{-4}% haematein solution	Sn^{4+}		8 × 10^{-8}	232
CPE	Solochrome Violet RS solution	Al^{3+}			233
GE	0.1 M H$_2$SO$_4$ + 10^{-6} M triphenylmethane dyes (crystal violet, methyl violet, malachite green)	Sb^{3+}	CSV	3 × 10^{-8}	234
GE	1 M HCl + 6 × 10^{-4} M Rhodamine C	Sb^{3+}	CSV	5 × 10^{-8}	38
GE	1.5 M NH$_4$SCN + 0.03 M (C$_2$H$_5$)$_4$NI + 0.3 M (CH$_2$NH)$_2$	Co^{2+}	CSV	10^{-6}	38

62

Electrode	Solution	Ion	Method	Detection limit	Ref.
GE	0.01 M HCl + 0.05 M [$(C_2H_5O)_2PSS_2$]Ni	Sn^{4+}	CSV	2×10^{-7}	38
GE	0.1 M KCl solution saturated with cinnamic acid	Cr^{3+}	Derivative ASV	2×10^{-7}	38
GE	0.1 M NH_4Cl + 0.1 M NH_3 solution + 0.04% 1-nitroso-2 naphthol	Co^{2+}	CSV	1×10^{-8}	38
MFE	2.5×10^{-4} alkaline solution of dimethylglyoxime (pH 9–11)	Ni^{2+}	CSV	2×10^{-6} gl^{-1}	235
GCE	0.002 M $RuCl_3$ + 0.002 M $K_4Ru(CN)_6$ + 0.5 M NaCl (pH 2)	As^{3+}	ASV	5×10^{-6}	236
GE	0.1 M H_2SO_4 + 0.1 M KCl + 5×10^{-7} M methyl violet	I^-	CSV	4×10^{-9}	34
GE	1.5 M HCl + $(6 \times 10^{-4} - 1.25 \times 10^{-3})$ M $Hg(NO_3)_2$ + 0.4 M KI	As^{3+} Te^{4+}	CSV	2.5×10^{-9} 5×10^{-10}	34
HMDE	10^{-5} M 1,2-dihydroxyanthraquinone-3-sulfonic acid (pH 7.1–7.3)	Al^{3+}	CSV	1×10^{-9}	237
Hg/GE	0.1 M $H_2C_2O_4$ + 7×10^{-6} M $Hg(NO_3)_2$	Cu^{2+}	DPASV	2×10^{-10}	238
HMDE	0.01 M NaOH, 2×10^{-7} M 2-(5'-bromo-2'-pyridilazo)-5-diethylaminophenol	Bi^{3+}	d.c.SV	7.5×10^{-11}	239
GE	2×10^{-5} M diphenylcarbazone + ammonium-chloride buffer solution (pH 9.2)	Mn^{2+}	Derivative CSV	8×10^{-9}	216

As a rule, two types of mechanical treatment are used, that is, renewal of the electrode surface by cutting off a thin layer [123] or polishing with abrasives.

Of interest is the use of renewable metal electrodes in a flow-through cell [123]. The electrodes are rods made of hardened epoxy resin, with a wire of an appropriate metal (13 metals were tested) passing along the rod axis. The electrode working surface is renewed by cutting off a thin layer (2–8 μm) from the end face of the electrode rod. The construction gives signal reproducibility within $\pm 3\%$ under continuous operation (about 10^5 cycles of the surface renewal).

Thermal regeneration of the electrode surface is used too. A platinum electrode is heated to a temperature of 600 or 1100 °C [112]. Such a treatment makes the electrode catalytically active.

Glassy carbon is activated by heating to 3000 °C. A similar effect is achieved through chemical oxidation of the surface [244]. However, Mentus and co-workers [127] think that an electrode is most efficient when its surface is fresh. Mechanical treatment uncovers active centers for electron transfer, which are lost after chemisorption of oxygen. The electrode surface is adequately cleaned of surface polymer films and is properly activated if laser treatment is used [245, 128].

Electrochemical methods of treatment and regeneration of the electrode surface are most popular. At sufficiently high potentials the electrode surface is modified with oxidation products during electrochemical preparation [246, 247]. As a result, for example, a substantial rise is observed in sensitivity and resolving power of the determination of hydrazine and its derivatives [248] and copper [246] on carbon electrode. At sufficiently negative potentials the electrode surface is modified with reduction products of water, ions, or impurities specially introduced into the solution (modification with adatoms) [82].

Mercury and mercury-plated electrodes are polarized at potentials that are several dozens of millivolts lower than the mercury dissolution potential [69].

The use of scanning tunneling microscopy [249] for in situ characterization of electrochemically activated glassy carbon electrodes shows that frequent treatments of the electrodes at a constant potential $+2.0$ V for 3 min and -1.0 V for 1 min result in considerable growth of the surface roughness. As a consequence, measurement reproducibility is impaired and residual current increases. Treatment conducted under square-wave conditions at a frequency 1 Hz and an amplitude $+2.0/-1.0$ V for 10 min brings about little or no change in the surface roughness and ensures better reproducibility of measurement results.

A detailed description of methods used to prepare glassy carbon electrodes for voltammetry is given in [250, 251] and graphite electrodes for stripping voltammetry in [14].

In spite of the great experience accumulated in the application of different methods of electrode regeneration, this problem unfortunately connot be considered as completely solved [252].

3.6. DESIGNS OF ELECTRODES AND ELECTROLYZERS

Stationary and rotating electrodes are widely used. The former electrodes are shaped approximately as a sphere (mercury), a flat disk, or a cylinder (solid metals, carbon). The working surface of rotating electrodes usually represents a disk or, sometimes, a disk and a ring. The surface area of such electrodes as a rule equals 0.05–0.3 cm^2. Designs of novel ultramicroelectrodes (UME) made of platinum, carbon fiber, or gold have been proposed [253]. One of the advantages of the UME is invariance of the current strength with time. The electrodes are used to study various processes in bioelectrochemistry and biology in vivo and in vitro [254, 255]. The possibility of using UME at high rates of potential scan is discussed in [256]. Low capacity of the double layer and low voltage drop permit operation of UME at potential variation rates up to 20,000 V s^{-1}. The values of the limiting current at the UME are rather stable and depend little on the voltage drop in the electrolyte, factors which allow measurements to be made at low concentrations of the supporting electrolyte [256]; or use exotic media as nonaqueous solutions without supporting electrolyte, which should find a wide application in HPLC [253]. Since the UME diameter is usually much less than the diffusion layer thickness, stirring of the solution is not required in some cases either at the stage of deposition or at the stage of the deposit dissolution [146]. To enhance sensitivity of determination with UME, it is proposed to set up a system comprising a large number of microelectrodes [257, 253].

Sternitzke and McCreery [258] presented a new method for precision fabrication of microelectrodes on the surface to produce arrays of nearly identical individual electrodes with uniform spacing. This involves the use of a pulsed nitrogen laser focussed onto an electrode surface through a microscope objective. Disk-type active sites which behave as microelectrodes are produced on a prepassivated glassy carbon surface.

The use of a newly developed electrode for tensammetry is reported [259]. The test solution flows through a glassy carbon cylinder (auxiliary electrode) and then through a short cylinder of silver (reference electrode). After passing these two electrodes the solution flows directly into the mercury. Here, it forms bubbles which, on reaching a certain size, go upward. With a constant flow rate, the solution will continuously bubble through the mercury. As long as the solution bubble is connected with the solution flowing through the other electrodes, the inner surface of the bubble constitutes the active electrode surface. In some way the bubble-electrode may also be regarded as a reverse mercury dropping electrode. The electrode is of fundamental interest for flow-injection analysis (FIA) and high-performance chromatography (HPLC).

Electrolyzers. Cells used in stripping electroanalytical chemistry can be classified as follows: two- and three-electrode cells with stationary and rotating electrodes; circulating and flow-through cells; thin-layer cells, and microcells.

Usual polarographic and voltammetric cells used for routine analysis are described in sufficient detail in monographs [4, 5, 14, 15, 72, 111].

Each of the designs of the electrochemical cell is intended for solving certain tasks of stripping analysis. The electrolyzer design should provide for a strict mutual arrangement of electrodes, should ensure the possibility of stirring the solution at the preelectrolysis stage or rotating the working (indicator) electrode, and should envisage bubbling of an inert gas through the solution.

Stirring of the solution in electrochemical cells is possible with a rotating working electrode, a magnetic stirrer, or a stream of inert gas. The last method is significant in that it provides for simultaneous deaeration of the solution.

In this case, the use of circulating [260] and flow-through [15, 117, 261–265] cells is more efficient. Construction of a circulating electrolyzer is illustrated in Fig. 3.1. As the working electrode in a circulating cell use is made of, for example, an amalgamated silver ball secured on a tapered end of a glass capillary. The test solution circulates through the gap between the ball and the walls of a teflon plate. The linear rate of the electrolyte circulation is adjustable over wide limits by selecting an appropriate plate and by changing the diameter of the indicator electrode silver ball [260]. Application of circulating electrolyzers makes it possible to improve considerably determination sensitivity owing to intensification of the concentration process, to obtain results to adequate reproducibility, and to decrease the solution volume required for analysis.

Automation of measurements favors wide application of electrochemical flow-through cells (cells with flow-through electrodes) in stripping electroanalytical chemistry [266–268]. For example, a flow-through cell with a stationary gold disk electrode ensures lowering of the detection limit of arsenic to 4×10^{-10} M as determined by the SV method [267].

In [262] a high-rate (100–300 ml/min) flow-through cell is proposed for

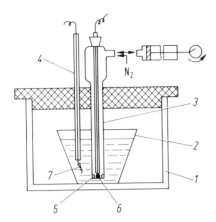

Figure 3.1. Circulating SV electrolyzer [260]; *1*, body; *2*, electrolyzer; *3*, working electrode; *4*, reference electrode; *5*, plate; *6*, amalgamated silver ball; *7*, platinum coil.

continuous determination of the concentrations of electroactive substances by SV method using a stationary mercury electrode. For determination of low Pb, Cd, and Zn concentrations, it is expedient to use a layer of compacted sodium sulphite (Na_2SO_3) to remove dissolved oxygen [262].

Use is made of cells with flow-through porous electrodes (FPE) and tube electrodes. Analyzers fitted out with FPE are characterized by a low detection limit because the large inner surface of the FPE washed by the moving phase ensures high efficiency of the analyte electrolysis [269]. A technology has been currently developed for the production of silver, silver chloride, silver sulphite, amalgamated silver, glassy carbon, graphite, and teflon-platinum-carbon FPE [269]. Good prospects for analytical applications of the FPE made of meshed glassy carbon are shown [269].

Glassy carbon FPE offer some advantages over other FPE, namely, good electroconductivity, high chemical stability, and a homogeneous structure [269]. For the determination of heavy metals in natural waters, use is made of cells with a reticulated vitreous carbon flow-through electrode with a mercury film deposited on the electrode surface [270].

Tube electrodes for flow-through cells are made of platinum and graphite; glassy carbon is not used in these electrodes owing to the problems involved in its mechanical treatment. In spite of simplicity and excellent hydrodynamic properties, cells with tube electrodes have not found extensive application because of the difficulties connected with the treatment of the tube inner surface. Electrochemical cells are frequently used as electrochemical detectors in flow-injection analysis [117, 263, 264]. The cells are convenient since they can be used in conjunction with sample pretreatment units, for example, ion-exchange columns, for removal of interfering elements.

The volume of the test sample can be substantially reduced if thin-layer and microelectrochemical systems are used [15, 271, 272].

In these systems, a silver disk coated with a mercury film is used as the working electrode. The auxiliary electrode is a platinum or amalgamated silver disk positioned parallel to the working electrode at a distance of 10–500 μm. Thin-layer cells with membrane metal electrodes [273] and a cell modification with two indicator electrodes [272] (Fig. 3.2) are developed.

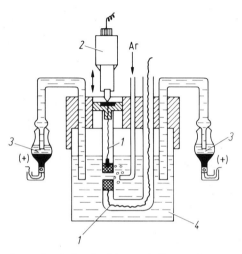

Figure 3.2. Thin-layer electrochemical cell with two indicator electrodes [272]: *1*, indicator electrodes with silver disks; *2*, electrode gap adjusting micrometer; *3*, reference electrodes (s.c.e.); *4*, glass vessel.

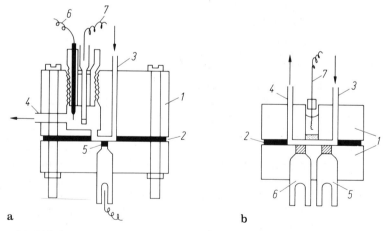

Figure 3.3. Thin-layer electrochemical detectors with divided (*a*) and undivided (*b*) electrode space [274]: *1*, detector body; *2*, gasket; *3*, detector inlet; *4*, detector outlet; *5*, working electrodes; *6*, auxiliary electrode; *7*, reference electrode.

A simple microelectrolytic cell [271] is suggested for voltammetric measurements in a single droplet of the test solution about 50 μl in volume. The working electrode is a disk CPE about 3 mm in diameter, with its working surface facing upward. A droplet of the test solution is applied to the working surface and the capillary (diameter 1.8 mm) of a silver-chloride reference electrode is immersed into the droplet. Thin-layer electrochemical systems find extensive application, similar to the flow-through systems, as electrochemical detectors specifically in high-perfomance liquid chromatography [274] (Fig. 3.3).

Compact electrochemical probes containing working, reference, and counter electrodes with diameters in the micrometer range have been prepared [275]. These multimicroelectrode devices can be immersed in a drop of solution, or presumably may be used as a miniature detectors. They may also be used in stripping analysis.

CHAPTER

4

STRIPPING ELECTROANALYTICAL METHODS IN THE ANALYSIS OF SOLUTIONS

4.1. ELECTROCHEMICAL CONCENTRATION

Any variant of the SEAM used to analyze solutions includes the stage of preliminary concentration (accumulation) of the solution component to be determined (analyte) on the surface or in the body of the working electrode. The concentrate formation processes can be divided into two main groups: processes involving charge transport (electrochemical or redox chemical reaction), and those involving no changes in the degree of oxidation of the analyte (e.g., adsorptive concentration).

The composition of the concentrate produced is the dominant factor that determines the registered analytical signal (response). The principal versions of preconcentration procedures and the types of the electrode processes used are considered in detail in [14, 15, 38]. Let us briefly examine some of them.

The first type includes the processes of metal discharge-ionization at the surface of a mercury or solid electrode considered in Chapters 1 and 2.

For concentration, it is essential to choose correctly the preliminary electrolysis potential. As a general approach to selection of this parameter, it is reasonable to recommend the analysis of curves in the coordinates "maximum current of the concentrate electrochemical transformation—electrolysis potential" exhibiting a characteristic shape of the polarographic wave (Fig. 4.1).

Tables 4.1 and 4.2 furnish basic electrochemical characteristics of the discharge-ionization of some metals on mercury and graphite electrodes in various supporting electrolytes. The characteristics assist in selecting conditions of the metal determination and also permit drawing a conclusion on the possibility of direct determination of a particular element in the presence of high amounts of an accompanying element. For example, the values of the discharge and ionization potentials of Cd, Pb, Cu, and Zn suggest that the first three elements can be determined in the presence of practically any amount of zinc with 1 M KCl as the supporting electrolyte ($E_{el} = -0.9$ V)—this was actually the case in analyzing standard reference samples of zinc. The effect of the matrix in determining Cd, In, Pb, and Cu in aluminium was eliminated by analogous selection of the electrolysis potential.

71

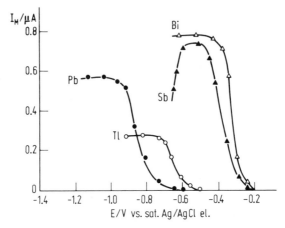

Figure 4.1. Maximum ionization currents of Bi, Sb, Pb, and Tl versus preelectrolysis potential (pseudopolarograms). (Indicator electrode: Bi-IGE; Sb-SMDE; Pb-GCE; Tl-Hg/CE. Supporting electrolyte: Bi and Sb—1 M HCl; Pb—0.2 M KBr; Tl—1 M NH$_4$OH + 1 M NH$_4$Cl; τ_{el} = 2 min; v = 1 V/min; c^0 = 2 × 10^{-7} M).

The second type of reaction is represented by the processes of oxidation or reduction of variable-valence ions accompanied by a chemical reaction leading to formation of a low-soluble compound on the electrode surface. The compound thus formed is subsequently reduced or oxidized electrochemically and the current arising therewith is measured. In the simplest case, these processes can be presented schematically as follows:

$$M^{n+} \rightarrow M^{(n\pm m)+} \pm me^- \tag{4.1}$$

$$M^{(n\pm m)+} + (n \pm m)A^- \rightarrow MA_{n\pm m}\downarrow \tag{4.2}$$

$$MA_{(n\pm m)} \pm me^- \rightleftarrows M^{n+} + (n \pm m)A^- \tag{4.3}$$

where M^{n+} and $M^{(n\pm m)+}$ denote an element with different degrees of oxidation; A^- is the solution component (hydroxide-ion, inorganic, or organic reagent) forming a low-soluble compound with the products of the electrode reaction (Eq. 4.1); for simplicity, the component A is given as a monovalent anion.

The sequence of the stages (Eqs. 4.1–4.3) written as E–C–E (electrochemical–chemical–electrochemical) underlies the stripping voltammetry of variable-valence ions. The optimum conditions of determining ions of some elements with the use of processes (Eqs. 4.1–4.3) are presented in Tables 4.3 and 4.4.

Table 4.1. Amalgam-Voltammetric Characteristics of Some Metals in 0.1 M Solutions of Supporting Electrolytes

($V_{el} = 5$ ml; $\tau_d = 3$ min; $v = 400$ mV/s; $c_i = 3.10^{-8}$ g ml^{-1}) [276][a]

Metal Ion	NaCl			(NH₄)₂SO₄			KOH			HCl			HNO₃			H₃PO₄			H₂C₂O₄		
	E_p	$I \times 10^8$	σ_{hp}	E_p	$I \times 10^8$	σ_{hp}	E_p	$I \times 10^8$	σ_{hp}	E_p	$I \times 10^8$	σ_{hp}	E_p	$I \times 10^8$	σ_{hp}	E_p	$I \times 10^8$	σ_{hp}	E_p	$I \times 10^8$	σ_{hp}
Zn²⁺	−0.97	8.0	50	−1.00	2.0	50	−1.07	26	40	−0.91	7.5	40	—	—	—	—	—	—	—	—	—
Cd²⁺	−0.61	7.0	35	−0.59	3.3	35	−0.63	7.5	30	−0.64	8.5	30	−0.62	6.0	30	0.53	2.5	50	−0.60	3.2	47
In³⁺	−0.46	5.6	7	−0.43	0.6	17	−0.66	8.0	17	−0.59	0.13	16	−0.50	10.3	17	—	—	—	−0.65	1.4	30
Sn⁴⁺	−0.44	6.0	15	−0.44	1.5	15	−0.44	0.8	17	−0.49	1.0	13	−0.49	0.2	12	—	—	—	−0.60	0.2	18
Tl⁺	−0.50	20.0	150	−0.47	6.0	120	−0.51	13	90	−0.54	13.5	127	−0.58	14.0	110	−0.55	13.0	120	−0.58	13.0	110
Pb²⁺	−0.41	6.5	35	−0.40	2.7	40	−0.49	7.7	33	−0.44	10.0	30	−0.46	6.0	30	−0.38	6.5	45	−0.47	6.7	55
Sb³⁺	−0.22	18.0	30	−0.18	1.4	17	−0.29	5.2	33	−0.18	7.3	28	−0.14	7.0	27	−0.12	8.0	47	−0.22	6.6	33
Cu²⁺	−0.12	3.5	52	+0.1	0.6	50	−0.05	15	35	−0.16	5.0	57	−0.005	8.0	40	+0.04	8.0	65	−0.10	11.5	37
				10.06	2.6																

Metal Ion	NH₄F			Na₄P₂O₇			LiOH			HBr			H₂SO₄			HClO₄			0.1 M NH₄F in 10% C₂H₅OH		
	E_p	$I \times 10^8$	σ_{hp}	E_p	$I \times 10^8$	σ_{hp}	E_p	$I \times 10^8$	σ_{hp}	E_p	$I \times 10^8$	σ_{hp}	E_p	$I \times 10^8$	σ_{hp}	E_p	$I \times 10^8$	σ_{hp}	E_p	$I \times 10^8$	σ_{hp}
Zn²⁺	−1.02	13	40	—	—	—	−1.10	8.0	77	—	—	—	—	—	—	—	—	—	−1.02	17.0	45
Cd²⁺	−0.62	8.0	30	−0.64	3.7	55	−0.65	4.8	33	−0.70	12.5	30	−0.56	4.5	40	−0.63	5.5	37	−0.60	12.0	30
In²⁺	−0.60	13	23	—	—	—	−0.84	2.2	13	−0.62	0.1	17	—	—	—	—	—	—	−0.51	1.6	17
Sn⁴⁺	—	—	—	—	—	—	—	—	—	−0.51	0.4	15	—	—	—	—	—	—	—	—	—
Tl⁺	−0.51	24	93	−0.56	6.7	130	−0.43	12	120	−0.59	13.0	110	−0.55	12.5	110	−0.56	13.0	105	−0.58	12.0	80
Pb²⁺	−0.43	9.0	30	−0.52	7.7	65	−0.55	4.7	42	−0.48	12.5	32	−0.40	6.5	35	−0.45	7.2	37	−0.44	14.0	30
Sb³⁺	−0.05	9.0	14	−0.17	0.4	23	−0.35	4.8	40	−0.17	11.7	33	−0.13	6.3	27	+0.08	11.7	40	−0.18	0.6	13
Cu²⁺	−0.07	6.0	47	−0.15	4.0	45	−0.18	5.3	37	—	—	—	+0.015	13.3	30	+0.03	7.1	44	−0.09	4.7	55

[a]Note that E_p are given relative s.c.e.; σ_{hp} is the anodic half-peak width.

73

Table 4.2. Voltammetric Characteristics[a] of Discharge-Ionization of Metals at Graphite Electrodes in some Electrolytes
$(c_0 = 5 \times 10^{-7}\,M; \; v = 1\,V/min; \; \tau_{el} = 2\,min)$ [14, 38]

Ion	Supporting Electrolyte	$E_{1/2}/V$	E_{el}/V	E_p/V
Cu^{2+}	1 M KNO_3	+0.35	−0.6	−0.1
	1 M KSCN (pH 2)	−0.3	−0.7	−0.25
	0.02 M tartaric acid + 0.02 M	—	−0.8	−0.4
	NH_4Cl + NaOH (pH 9)			
	1 M NaSCN + 1 M KOH	—	−1.3	−0.4
	1 M KOH + 0.1 M tartaric acid	—	−1.6	−0.5
	1 M KOH + 0.1 M ethylenediamine	—	−1.2	−0.5
Ag^+	1 M KNO_3	+0.05	−0.4	+0.1
	1 M KSCN	+0.3	−0.6	−0.15
	0.2 M H_2SO_4	+0.02	−0.6	+0.15
Au^{3+}	1 M HCl	+0.5	−0.2	+0.5
	1 M HNO_3	+0.75	−0.2	+0.8
	1 M KSCN (pH 8)	+0.4	−0.6	+0.4
Zn^{2+}	1 M KOH + 0.1 M ethylenediamine	—	−1.6	−1.3
	1 M KSCN (pH 5)	−1.02	−1.3	−1.05
	1 M KSCN (pH 6)	−1.1	−1.6	−1.1
Cd^{2+}	0.1 M HCl	−0.72	−1.0	−0.75
	1 M KCl (pH 2)	−0.78	−1.0	−0.72
	1 M KSCN	—	−1.2	−0.78
	1 M KNO_3 (pH 3)	—	−1.1	−0.64
	1 M NH_4Cl + 1 M NH_4OH	−0.81	−1.1	−0.82
Hg^{2+}	0.1 M KNO_3	−0.1	−0.4	+0.1
	1 M KSCN	−0.25	−0.6	−0.12
In^{3+}	0.1 M KCl (pH 3)	−0.9	−1.4	−0.72
	1 M KBr	−0.8	−1.2	−0.76
	1 M NH_4SCN	−0.9	−1.3	−0.70
Tl^+	0.2 M NH_4OH + 0.1 M NH_4Cl (pH 8)	−0.95	−1.4	−0.48
	0.1 M NaSCN	—	−1.2	−0.47
	0.1 M KCl	—	−1.2	−0.47
Sn^{2+}	2M HCl	−0.6	−1.1	−0.55
	1 M CH_3COOH + 1 M CH_3COONa	—	−1.0	−0.68
	1 M $C_2H_2O_4$	—	−1.0	−0.48
Pb^{2+}	0.1 M KNO_3	−0.4	−0.8	−0.43
	0.1 M HCl	−0.48	−1.0	−0.55
	1 M CH_3COOH + 1 M CH_3COONa	—	−1.0	−0.48
	1 M NH_4Cl + 20-% citric acid	—	−0.85	−0.55
Sb^{3+}	1 M HCl	−0.25	−0.6	−0.22
Bi^{3+}	1 M NH_4Cl	−0.2	−0.5	−0.15
	0.1 M HCl (pH 1)	−0.2	−0.4	−0.08
	1 M KOH + 0.25 M ethylenediamine	—	−0.6	−0.18

Table 4.2. (*Continued*)

Ion	Supporting Electrolyte	$E_{1/2}/V$	E_{el}/V	E_p/V
Te^{4+}	1 M HCl	—	−0.6	+0.37
	1 M KBr (pH 2)	—	−0.6	+0.25
Fe^{3+}	1 M KOH + 0.01 M tartaric acid	−1.4	−1.7	−0.8
	0.05 M Na tartrate (pH 5)	—	−1.6	−0.6
Co^{2+}	0.2 M NH_4OH	—	−1.2	−0.4
	0.1 M K_2SO_4	—	−1.4	−0.1
	1 M KOH + 0.1 M tartaric acid	—	−1.6	−0.6
	0.1 M KSCN	—	−1.2	−0.5
Ni^{2+}	0.1 M NH_4Cl + 0.1 M NH_4OH	−1.1	−1.2	−0.5
	1 M KNO_3 + 0.001 M HNO_3	−1.1	−1.2	−0.1
	0.1 M KSCN	—	−1.2	−0.5
Pd^{2+}	1 M H_2SO_4	—	−0.6	+0.4
	1 M HCl + 1 M KCl	—	−0.6	+0.4
	1 M $HClO_4$	—	−0.4	+0.35

[a] Potentials are given with reference to s.c.e.

The third and fourth types of electrode processes comprise an electrochemical [38] or chemical [277, 278] reaction of the analyte ions with the electrode material or an auxiliary element. The determination procedure includes the E–C–E or C–E stages. Thus, the ability of halogenide-ions (Cl^-, Br^-, I^-) and S^{2-}, CrO_4^{2-}, VO_3^-, WO_4^{2-}, and MoO_4^{2-} ions to form low-soluble compounds with the electrode metal ions (Hg, Ag, etc.) provides the basis for the third type of the element concentration in the SEAM, which is called [38] the stripping voltammetry of anions. At present, the use of this version for the determination of anions is limited, since there exist enough methods for determining the abovementioned anions with a lower time expenditure compared to SEAM, for example, with the use of ion-selective electrodes. However, interaction of the analyte with the electrode material is used to advantage for determining the concentration of some organic compounds.

The concentration process may rest on chemical redox reactions between the analyte and the electrode material, resulting in the formation of low-soluble compounds or amalgams on the electrode surface. Subsequently, the compounds or amalgams are dissolved electrochemically. In many cases this method of concentration enables one to exclude the stages where the analyte and matrix elements are separated and to simplify the analysis. A strict fulfilment of the conditions found for this or that chemical reaction proceeding at the electrode provides for a higher selectivity of the preconcentration stage

Table 4.3. Determination of Elements in the Form of Low-Soluble Oxides, Hydroxides, and Salts by SEAM [14, 38][a]

Analyte	Electrode Reactions	Electrolyte Composition	$E_{1/2}$/V	E_{el}/V	E_p/V	$C_{min} \times 10^7$ M
Tl	$Tl^+ + 3OH^- \rightleftarrows Tl(OH)_3 + 2e$	0.35 M $(NH_4)_2SO_4$ + NH_4OH (pH 7)	+1.0	+1.2	+0.3	5
Ce	$Ce^{3+} + 4H_2O \rightleftarrows Ce(OH)_4 + e$	0.1 M CH_3COOH + 0.1 M CH_3COONa	+0.7	+1.0	+0.3	100
Pb	$Pb^{2+} + 2H_2O \rightleftarrows PbO_2 + 4H^+ + 2e$	Buffer solutions (pH 1–13)	$-1.4-$	$-1.6-$	$-0.8-$	2
			-0.5	-0.7	0.0	
Cr	$CrO_4^{2-} + 4H_2O + 3e \rightleftarrows Cr(OH)_3 + 5OH^-$	0.4 M NH_4Cl + 0.1 M NH_4OH	-0.45	-0.7	+0.6	0.4
Mo	$MoO_4^{2-} + 4H^+ + 2e \rightleftarrows MoO_2 + 2H_2O$	0.5 M $(NH_4)_2SO_4$ + 0.05 M EDTA	—	-1.4	-0.15	0.3
Mn	$Mn^{2+} + 4OH^- \rightleftarrows Mn(OH)_4 + 2e$	2 M $(NH_4)_2SO_4$ + H_2SO_4 (pH 5)	+0.75	+0.9	+0.25	5
Mn	$Mn^{2+} + 4IO_4^+ \rightleftarrows Mn(IO_4)_4 + 2e$	$(0.02–0.2)$M HNO_3 + 3×10^{-2} M KIO_4	+1.3	+1.4	+1.1	4
Re	$ReO_4^- + 4H^+ + 3e \rightleftarrows ReO_2 + 2H_2O$	$(4–5)$M H_3PO_4	-0.8	-0.9	-0.25	0.1
Fe	$Fe + 3OH^- \rightleftarrows Fe(OH)_3 + e$	H_3BO_3 + NaOH (pH 8)	—	-0.05	-0.5	8
Fe	$Fe^{3+} + 2OH^- \rightleftarrows Fe(OH)_2 - e$	0.05 M citric acid + NaOH (pH 10)	—	-1.0	-0.7	20

[a]Potentials are given relative s.c.e.

Table 4.4. Determination of Metal Ions in the Form of Low-Soluble Compounds with Organic Reagents by SEAM [14, 35, 38][a]

Analyte	Electrode Reactions	Electrolyte Composition	E_{el}/V	E_p/V	$c_{min} \times 10^8$/M
Ce	$Ce^{3+} + Ft \rightleftarrows [Ce^{4+} Ft] + e$	(0.1–0.3) M HNO_3 + 0.1% Ft	+1.2	+0.8	0.5
Sb	$[SbCl_6]^{3-} + Rd^+ \rightleftarrows Rd\,SbCl_6 + 2e$	1.5 M KCl + 0.5 M H_2SO_4 + 4×10^{-4} M Rd^+	+0.8	+0.35	4
Sb	$[Sb(OH)_n Cl_{6-n}]^{3-} + Kt^+ \rightleftarrows KtSb(OH)_n \cdot Cl_{6-n} + 2e$	1.2 M HCl + 5×10^{-4} M Kt^+	+0.8	+0.4	0.2
I	$2I^- + Cl^- + Rd^+ \rightleftarrows RdI_2Cl$	0.1 M H_2SO_4 + 0.1 M KCl + 5×10^{-5} M Rd^+	+0.8	+0.35	0.5
Co	$Co^{2+} + 3DDTK \rightleftarrows Co(DDTK) + e$	0.5 M NH_4OH + 0.5 M NH_4Cl + 5×10^{-5} M $\cdot$ NaDDTK	−0.5	−1.1	5
Ni	$Ni(HD)^+ + 2OH^- \rightleftarrows Ni(OH)_2HD + e$	0.03 M KOH + 1×10^{-6} M HD	+0.8	+0.4	0.2

[a]DDTK—diethyldithiocarbaminate ion; Rd—rhodamine cation; Kt—triphenylmethane dye cation; Ft—phytin; HD—dimethylglyoxime; potentials are given relative s.c.e.

[279]. To conclude this consideration of the concentration of elements by the SEAM in the form of low-soluble compounds with inorganic and organic compounds, it is worth emphasizing that the potential of this variant is tremendous but it is used insufficiently. In future we expect more intensive theoretical advances and practical solutions of analytical problems on the basis of highly selective electrochemical reactions along with the use of highly selective organic and inorganic precipitants.

4.2. ADSORPTIVE CONCENTRATION

In a recent paper [280], a question is posed: "Polarographic Adsorption Analysis and Tensammetry: Toys or Tools for Day-to-day Routine Analysis?" We consider adsorptive concentration to be more a "tool" than a "toy."

Adsorptive concentration has become popular in recent years [281]. This variant of the SEAM allows the determination of alkaline-earth and rare-earth metals, aluminum, vanadium, cobalt, and nickel, to say nothing of the elements traditional for the SEAM, such as Sn, Pb, Cd, and others. Besides, adsorptive concentration (A) broad prospects open up for the determination of organic compounds by the SEAM (processes types A–C–E, C–A–E, A–E). Adsorptive concentration of metals is associated as a rule with the use of surfactants added to the solution in abundance. It is essential that the potentials of the reagent electrochemical transformations and of its compound with metal vary as much as possible. Adsorptive concentration is advantageous in its high selectivity of determination and the possibility in some instances of concentration carried out with the open circuit. Some examples of the AdSV are given in a monograph by Wang [29].

4.2.1. Adsorptive Concentration of Metal Complexes

If in the test solution the metal ion M^{n+} forms with a ligand L^- (for simplicity, assign a unity negative charge to the ligand) a complex M_pL_q (p and q being stoichiometric coefficients) possessing surface-active properties, the complex can be concentrated on the working electrode surface. Several mechanisms of such concentration are possible, including an adsorption stage.

1. In the simplest case, concentration of M^{n+} can be described by Eqs. 4.4 and 4.5.

$$pM^{n+} + qL^- \rightleftarrows [M_pL_q]_{(sol)}^{pn-q} \tag{4.4}$$

Reaction 4.4 proceeds in the solution (sol) and represents the chemical stage

(C) of the process. The complex formed adsorbs (ads) on the electrode

$$[M_pL_q]_{\text{sol}}^{pn-q} \rightleftarrows [M_pL_q]_{\text{(ads)}}^{pn-q} \qquad (4.5)$$

The general scheme of the process can be written as C–A. The amount of adsorbate on the electrode depends on the concentration of M^{n+} in the test solution and on the time of contact between the working electrode and the solution. Usually, the concentration relation $c_L \gg c_{M^{n+}}$ is maintained to ensure a stable composition of the complex formed.

2. Another mechanism of adsorbate formation is possible. The ligand is adsorbed initially on the electrode (A)

$$qL_{\text{(sol)}}^- \rightleftarrows qL_{\text{(ads)}}^- \qquad (4.6)$$

At the second stage (C) adsorbed ligands interact with M^{n+} ions

$$pM^{n+} + qL_{\text{(ads)}}^- \rightleftarrows [M_pL_q]_{\text{(ads)}}^{np-q} \qquad (4.7)$$

In this case, as distinct from Reaction 4.4, the Reaction 4.7 proceeds at the electrode surface. The sequence A–C is true for the given mechanism which is realized at electrodes modified in situ (Chapter 3). The amount of adsorbate on the electrode depends on the concentration of M^{n+} or, more accurately, on the amount of M^{n+} that reacted with adsorbed L^- ions whose adsorbability on the given electrode should be as high as possible.

If the rates of the stages A and C are comparable, adsorption of the ligand and formation of the complex on the electrode will proceed almost simultaneously (in situ) and it will be difficult to distinguish between them.

3. A more complex situation is not altogether impossible. Let M^{n+} form no surface-active complex with L^-, but the product of the electrochemical transformation of M^{n+} forms such a complex. Then it is possible to concentrate M after electrooxidation or electroreduction of M^{n+} (stage E)

$$pM^{n+} \pm me^- \rightarrow pM^{n\pm m} \qquad (4.8)$$

$$pM^{n\pm m} + qL^- \rightleftarrows [M_pL_q]_{\text{(ads)}}^{p(n\pm m)-q} \qquad (4.9)$$

When M is concentrated following this mechanism, the sequence of stages can be represented as E–C–A. It is easy to show that the amount of $[M_pL_q]^{p(n\pm m)-q}$ adsorbed on the electrode is determined by the concentration of M^{n+} in the test solution.

4. Finally, concentration of M frequently proceeds following a mechanism that combines the second and third variants described above. In this case the sequence of stages is given as A–E–C.

An analysis of the set of possible reactions makes it possible to draw some conclusions that are important in practical applications of the SV:

(i) The adsorptive concentration of metals in the form of complexes can add to the SV selectivity, since the choice of some specific reagent (ligand) and a certain degree of oxidation of the metal to be determined can provide for the conditions of the metal determination when mixed with ions of other metals.

(ii) The use of a ligand capable of forming surface-active complexes with several components of the test solution provides for group adsorptive concentration of metals, although in this case one often encounters the problem of finding conditions for individual determination of the adsorbate components.

(iii) For systems that comprise successively A–C and A–E–C, stage A consists in surface modification in situ and therefore this stage can be replaced by preliminary modification of the electrode surface using an appropriate reagent. In this way adsorptive SV is bridged with electro-analysis employing chemically modified electrode.

(iv) The detection limit of the elements by the AdSV method is extremely low, because it is independent of the product of the compound solubility when concentrated either in the form of salts or as hydroxides; the upper concentration limit is determined by saturation of the electrode surface.

(v) In the case of adsorptive concentration comprising C–A and A–C, electrochemical properties (reduction potential) of the metals entering the composition of adsorbed complexes are not so decisive as for electrolytical concentration, a fact which gives grounds to expect broadening of the list of elements determined by the SV method.

Whatever the mechanism of the adsorptive concentration of metal complexes, the informative stage is electrochemical (E). However, this does not imply that the registered response is due to electrochemical processes that are similar in nature. In many cases, for example, when trace Al [233] is subject to adsorptive concentration as a complex with Solochromviolet RS, the response is due to the reduction of the adsorbed complex. Here the reagent itself is also adsorbed on the electrode. However, the reduction potential of the reagent (in our case it is 0.3 V lower) differs from the reduction potential of the complex. This makes it possible to resolve currents of adsorbed complexes and the reagent in voltammograms.

Studies of some systems, for example, Sb(III) with triphenylmethane dyes [234], show that the stage of adsorption of the dye molecules precedes electro-oxidation of Sb(III) to Sb(V) with subsequent formation of a low-soluble compound, that is, the concentration proceeds following the adsorption mechanism A–E–C. In such systems the response appears as a result of an electrochemical reverse-transition reaction, like transition of antimony to the trivalent state. The same mechanism is involved in the concentration and determination of Cr(VI) with diphenylcarbazide [282].

Another possible way of generating responses in the AdSV of metal complexes are hydrogen catalytic currents due to catalytic activity of the adsorbate. Systems of this type that are best known is AdSV are Ni and Co dimethyl-glyoximates [283, 284] which have become classical examples.

Thus, AdSV is not only has a large potential at the accumulation stage but also allows the choice of the method that is most convenient for generating responses under particular conditions.

Table 4.5 specifies conditions of determining concentration of ions of some metals by AdSV. As can be seen, the AdSV allows the determination of approximately 30 elements including those (Th, Tc, Zr, Tb, Y, Al, etc.) that cannot be determined by other SV methods. At the same time, AdSV is applicable to selective determination of some metals (Zn, Cu, Sn, Au, Ga, Pd) which can be also concentrated in the elemental state using electrolysis. In some instances (alkaline and alkaline-earth metals, rare-earth elements), adsorptive concentration is of group character and, as follows from the E_p values (Table 4.5), it is impossible to resolve response of the above elements only by the SV means even with high resolving power of modern instruments. Therefore, when analyzing mixtures of alkaline-earth metals [285] or rare-earth elements, one needs to perform preliminary chromatographic separation of the components. Estimates of the detection limits (Table 4.5) of ions of many metals are usually lower than for other SV variants (e.g., for Cu and Sn) [286]. However, none of the papers cited in Table 4.5 gives a rigorous statistical support of the given estimates and therefore it is quite possible that if the conditions of Table 4.5 are reproduced, the estimates of the detection limits will vary. Of importance is the fact that the data of Table 4.5 tell the analysis about the possibility of determining trace amounts of metals.

4.2.2. Adsorptive Concentration of Organic Compounds

At present, ample material has been accumulated on transformation of molecules of organic compounds, including adsorptive processes on the surface of mercury, platinum, graphite, and other electrodes. Information about studies made before 1973 can be found in the monograph by Jering [313], which systematizes data on adsorption of several hundred organic substances on mercury electrodes. The role played by the benzene ring in adsorptive and electrochemical transformations of aromatic compounds on platinum was considered by Kazarinov [314]. Processes taking place at mercury electrodes were considered by Frumkin [315].

Table 4.5. Determination of Metals Ions as Complexes with Organic Reagents by AdSV

Ion	Reagent	Supporting Electrolyte	Electrode	Mode	E_{el}, V (s.c.e.)	E_p, V (s.c.e.)	$c_{min} \times 10^9$, mol l^{-1}	Ref.
Na$^+$	Solochrome violet RS (1 × 10^{-6} M)	Acetic buffer solution (pH 4.7)	HMDE	DPCSV	-0.6	-0.98	1.8	285
K$^+$	Solochrome violet RS (1 × 10^{-6} M)	Acetic buffer solution (pH 4.7)	HMDE	DPCSV	-0.6	-0.95	2.0	285
Rb$^+$	Solochrome violet RS (1 × 10^{-6} M)	Acetic buffer solution (pH 4.7)	HMDE	DPCSV	-0.6	-1.01	0.8	285
Cs$^+$	Solochrome violet RS (1 × 10^{-6} M)	Acetic buffer solution (pH 4.7)	HMDE	DPCSV	-0.6	-1.03	2.0	285
Cu^{2+}	Cysteine (2 × 10^{-4} M)	0.5 M NaCl (pH 3.0)	HMDE	DPCSV	-0.5	-0.92	60.0	286
Au^{3+}	Dithizon (0.01 M in CHCl$_3$)	0.01 M HCl	CPE	a.c. SV	-0.1	-0.34	20.0	225
Zn^{2+}	Bromazepam (1 × 10^{-4} M)	Phosphatic buffer solution (pH 5–6)	HMDE	DPCSV	-0.5	-1.2	100.0	287
Al^{3+}	Solochrome violet RS (1 × 10^{-6} M)	0.2 M Acetic buffer solution (pH 4.5)	HMDE	DPCSV	-0.45	-0.61	55.0	233
Ga^{3+}	Solochrome violet RS (1 × 10^{-6} M)	0.2 M Acetic buffer solution (pH 4.5)	HMDE	DPCSV	-0.40	-0.56	63.0	288
La^{3+}	o-Crezophtalexone (1 × 10^{-4} M)	0.1 M NH$_4$OH + 0.1 M NH$_4$Cl (pH 9,4)	HMDE	DPCSV	-0.40	-1.10	1.2	230
Ce^{3+}	o-Crezophtalexone (1 × 10^{-4} M)	0.1 M NH$_4$OH + 0.1 M NH$_4$Cl (pH 9,4)	HMDE	DPCSV	-0.40	-1.12	1.7	230
Pr^{3+}	o-Crezophtalexone (1 × 10^{-4} M)	0.1 M NH$_4$OH + 0.1 M NH$_4$Cl (pH 9,4)	HMDE	DPCSV	-0.40	-1.08	1.4	230
Tb^{3+}	Cupferron (0.08%)	0.1 M NH$_4$OH + 0.3 M NH$_4$Cl + 0.05% dodecyltrimethylammonium	HMDE	DPCSV	-0.10	-0.70	4.4	289
Sm^{3+}	Cupferron (0.05%)	0.05 M NH$_4$Cl (pH 4.8)	HMDE	DPCSV	-0.30	-1.21	60.0	290
Ti^{4+}	Solochrome violet RS (1 × 10^{-6} M)	Phthalatic buffer solution (pH 1.5)	HMDE	DPCSV	-0.40	-0.92	0.7	291
Ti^{4+}	Cupferron (0.03 M)	KCl + HClO$_4$(HCl) (pH 3.0)	HMDE	DPCSV	-0.20	-0.81	200.0	292
Sn^{4-}	Cysteine (2 × 10^{-4} M)	0.4 M NaNO$_3$ (pH 2.8)	HMDE	a.c. SV	-0.5	-1.10	2.0	286
Sn^{4+}	Tropolone (4 × 10^{-6} M)	Acetic buffer solution (pH 4.0)	HMDE	DPCSV	-0.40	-0.65	2.4	293
Zr^{4+}	Solochrome violet RS (1.5 × 10^{-6} M)	Acetic buffer solution (pH 4.6)	HMDE	DPCSV	-0.30	-0.48	23.0	294
Cu$^+$	2,9-Dimethyl phenantroline	0.5 M NaCH$_3$COO (pH 6)	CPE	DPSV	Open circuit	0.26		295

82

					Open circuit			
Pb²⁺	Crown ethers	0.1 M Na₂CO₃	CPE	DPSV	−0.9	−0.66		296
Cr⁶⁺	Diethylenetriaminepenta-acetic acid (2 × 10⁻⁵ M)	2.5 M NaNO₃ (pH 6.2)	HMDE	DPCSV		−1.15	40.0	297
Cr⁶⁺	Diethylenetetraminepentacetic acid	LiNO₃ (pH 6.6)	HMDE	DPSV	−1.0	−1.35	20.0	298
Cr⁶⁺	Diphenylcarbazide (2 × 10⁻⁶ M)	0.1 M H₂SO₄	Ge	d.c.SV	+0.4	−0.32	20.0	282
Mo⁶⁺	7-Nitro-8-hydroxyquinoline-5-sulfonic acid	0.7 M HCl + 0.1 M H₂SO₄	HMDE	DPCSV	−0.2	−0.95	0.3	299
In³⁺	Oxine (1 × 10⁻⁴ M)	HAc + NH₄Ac (pH 4.0)	HMDE	d.c.SV	−0.72 → −0.5	−0.62	0.15	300
Mo⁶⁺	Caprolactam (0.2 M)	1 M HCl + 0.03 M KSCN	IGE	d.c.SV	−0.2	+1.10	12.0	301
Mo⁶⁺	Antipyrine (0.05 M)	0.5 M HCl + 0.03 M KSCN	IGE	d.c.SV	−0.3	+0.10	0.3	302
Mo⁶⁺	2-Benzylideneiminobenzohydroxamic acid (5 × 10⁻³ M)	Phosphate buffer solution (pH 3.3) + NaClO₄	HMDE	DPCSV	0.00	−0.26	1.0	303
UO₂²⁺	Xylidyl blue I (2 × 10⁻⁵ M)	1% Ethylenediamine + 0.02 M HCl (pH 10.5)	HMDE	DPCSV	−0.30	−0.86	30.0	304
UO₂²⁺	PAR (1 × 10⁻⁷ M)	Boratic buffer solution (pH 8.8–9.8)	HMDE	DPCSV	−0.4	−0.65	8.0	305
Th⁴⁺	Thenoyltrifluoroacetone (0.002 M)	0.1 M Acetic buffer solution (pH 5.2)	HMDE	DPCSV	−0.1	−0.62	10.0	306
Fe³⁺	3,5-Brom-PADAT (2 × 10⁻⁶ M)	Acetic buffer solution (pH 4.7)	HMDE	a.c.SV	−0.05	−0.43	24.0	307
Fe³⁺	Solochrome violet RS (5 × 10⁻⁶ M)	0.1 M Acetic buffer solution (pH 5.1)	HMDE	DPCSV	−0.40	−0.71	1.5	308
Ni²⁺	Dimethylglyoxime (0.02% in ethanol)	0.1 M Triethylamine + 0.2 M NH₄Cl	MFE	d.c.SV (DPCSV)	−0.75	−1.04	0.8	309
Co²⁺	Dimethylglyoxime (0.02% in ethanol)	0.1 M Triethylamine + 0.2 M NH₄Cl	MFE	d.c.SV	−0.75	−0.94	1.2	309
Co²⁺	2,2'-Dipyridyl (10⁻⁵ M)	0.025 M NH₄OH + 0.2 M NH₄Cl (pH 8.5–9.0)	HMDE	DPCSV	−1.2	−1.53	4.5	310
Co²⁺	Cyclohexane-1,2-dionedioxime	Buffer solution N-(2-oxyethyl-piperazine-N'-2-ethyl sulfonic acid + NaOH, pH 7.6	HMDE	DPCSV	−0.60	−0.87	0.6	311
Pd²⁺	Dimethylglyoxime (0.05 M)	Acetic buffer solution (pH 5.15)	HMDE	DPCSV	−0.20	−0.74	12.0	312

Electrochemical reactions of adsorbed organic substances which are characteristic of particular functional groups are well known [8, 9]. This is of special interest for the organic analysis of various classes of organic compounds.

Results concerning adsorption and faradaic transformation of mono- and oligonucleotides [316–319], nucleic acids, and their components [320, 321] are published. It is shown that alternating current (a.c.) voltammetry and recording of the out-of-phase component of the a.c. response [316–319], differential and normal pulse polarography (DPP and NPP) [320], linear sweep voltammetry (LSSV) [322] in conjunction with dropping mercury, and hanging mercury drop electrodes (HMDE) can give a more detailed insight into the real situation existing for adsorption of natural biopolymers at charged surfaces. This technique can also be used for the detection of small perturbations in the double-helical structure of the DNA induced by environmental agents [320, 323]; it can also be applied to the analysis of biopolymers [324], and in clinical [325] and pharmaceutical [326] chemistry.

The SV method enables us to obtain the same and additional information in the investigation and analysis of biologically important substances [30]. The pioneer works of Palecek [324, 327–331, 31] and Brabec [332] on stripping analysis of biopolymers serve as a proof of this.

However, the possibilities that have been realized represent possibly only a small portion of those that open up if we cast a glance at the results of studies of adsorption processes described in electrochemistry.

We can point out three ways of detecting adsorbed molecules on the electrode surface:

1. Recording of adsorption–desorption peaks (this may not be connected with the electrochemical transformations of the adsorbed substance) [313, 315, 316–319].

2. Recording of catalytic currents appearing in the presence of cobalt salts as a result of adsorption of proteins, for example [333, 334].

3. Recording of currents due to the electrochemical transformation of the adsorbed organic substances themselves [316–319, 321–323, 332, 335, 336] or of their compounds with the electrode material [324, 327–331].

In this book we shall consider in detail only the possibilities of those SEAM which include, as informative stages, processes of electrochemical transformation of the adsorbed organic substance to the investigated.

In recent years a large number of papers have been published which are dedicated to the use of adsorption for concentration and determination of organic substances by the SV method. Those papers are cited in the monograph by Wang [29].

Some examples of the determination of organic substances by the AdSV method with the use of different electrodes and techniques are given in Table 4.6.

It should be noted that AdSV becomes especially effective in combination with up-to-date methods of chromatography, because resolution of electrochemical responses of substances that are close in composition is not always sufficient for the identification of individual compounds. Strongly pronounced adsorbability of organic compounds allows their concentration from rather dilute solutions, thus ensuring an extremely low detection limit. The choice of an electrode material which shows affinity for the analyte improves determination selectivity and lowers the detection limit.

4.2.3. Adsorption from the Gaseous Phase

Sorption from the gaseous phase can be considered as a new method of concentration in SV [64]. The analyte is transformed by a chemical reaction to a volatile compound which is extracted from the solution using a gas stream. The gas stream is directed from a nozzle onto the surface of a disk electrode (wall-jet electrode). After a period of time the electrode is transferred to an electrolyzer and a voltammogram is registered. This method of concentration was used for the first time for the determination of mercury [65]. A well-known reaction of hydrogen sulphide and silver was used to determine concentration of S^{2-} ions [66]. Cl^-, Br^-, and I^- were oxidized in the solution using permanganate ions to the corresponding gases which were sorbed on a silver electrode [67].

The above-described method of concentration from the gaseous phase will probably prove to be rather useful in analysis of gases.

4.3. EXTRACTIVE CONCENTRATION AND ANALYSIS OF ORGANIC EXTRACTS

Extractive separation and concentration of trace elements are used extensively in SEAM. When the extraction stage is included in the analysis, of special interest is the possibility of the direct analysis of the extract. The applicability of this approach to the SEAM where mercury indicator electrodes are used was shown by Karbainov et al. [369], who developed the theory of the SV method for organic extracts. A study was made of the amalgam-voltammetric behavior of hydroxyquinolinate complexes of Cu, Pb, and Cd in mixtures (1:1) of ethanol with benzene, chloroform, and carbon tetrachloride, benzene with dimethylformamide, and diethyldithiocarbaminate complexes of the same metals in mixtues (1:1) of ethanol and benzene, chloroform, and carbon tetra-

Table 4.6. Determination of Organic Compounds by AdSV

Analyte	Supporting Electrolyte	Electrode	Mode	E_{el}, V (s.c.e.)	E_p, V (s.c.e.)	$c_{min} \times 10^9$, mol^{-1}	Ref.
Adenine	1 M H$_3$BO$_3$ + 0.4 M NaOH (pH 8.5) + 150 mM Cu^{2+}	HMDE	DPCSV	−0.15	−0.63	0.3	337
Adriamycin	Acetic buffer solution (pH 4.5)	CPE	CV and DPCSV	−0.50	−1.12	10.0	338
N-Alkylated anilines	0.1 M Phosphatic buffer solution (pH 7.4)	CPE	DPCSV	—	−0.76	50.0	339
Amethopterine	0.01 M Phosphatic buffer solution (pH 7.0)	SMDE	DPCSV	−0.30	−0.72	0.2	340
2-amino-5-sulpho-amoylnaphthalene-azo-(2-chloro-4'-nitrobenzene	Britton-Robinson buffer solution (pH 2.6) in 50% methanol	SMDE	a.c.SV	−0.25	−0.82	6.0	341
Atropine	0.05 M KOH	HMDE	DPCSV	−0.70	−1.35	50.0	281
Bromazepam	Acetic buffer solution (pH 5.0)	HMDE	DPCSV	−0.60	−0.92	10	342
Caprolactam	Acetic buffer solution (pH 4.5)	HMDE	DPCSV	0.0	−0.30	0.2	343
Catechol compounds	0.5 M KCl + 0.1 M HCl (pH 2) + 1 × 10^{-5} M Sn^{4+}	HMDE	d.c.SV	−0.10	−0.57	5.0	344
Chlordiazepoxide	Britton-Robinson buffer solution (pH 6.8)	HMDE	DPCSV	−0.65	−1.02	0.9	345
Codeine	1 M NaOH	HMDE	DPCSV	−0.70	−1.12	10.0	281
Cocaine	0.05 M NaOH	HMDE	DPCSV	−0.70	−1.20	10.0	281
Cyadox	0.05 M NaClO$_4$ in 5% DMFA	HMDE	DPCSV	+0.10	−0.52	10.0	346
Cysteine	Britton-Robinson buffer solution (pH 8.0)	MFE	a.c.SV	−0.05	−0.35	0.05	347
Daunorubicin	0.1 M Acetic buffer solution (pH 4.4)	CPE (HMDE)	DPCSV	−0.10	−0.47	1.0	348
Dibutyl tin chloride	1 M Acetic buffer (pH 4.2) solution	HMDE	DPASV	−1.20	−0.85	100.0	349
Dilthiazem	0.05 M NaOH	HMDE	DPCSV	−0.90	−1.55	4.0	350
N,N'-Dimethyl-4-amino-3'-methylazobenzene	Britton-Robinson buffer solution (pH 4.72) in 20% methanol	SMDE	DPCSV	−0.10	−0.57	20.0	351
Dodecylbensil sulphate	1 M NaOH	HMDE	DPCSV	−0.70	−1.20	10.0	281

Compound	Medium	Electrode	Technique				Ref.
Dopamine	0.1 M HCl	Pt	a.c.SV	−0.20	−0.45	50.0	352
EDTA	0.1 M acetic buffer solution (pH 5.2)	HMDE	DPCSV	−0.20	−0.54	30.0	353
Erythromycin	Phosphate buffer solution (pH 7.4)	HMDE	DPCSV	−1.0	−1.25	2.5	354
Folic acid	Britton-Robinson buffer solution (pH 5)	HMDE	DPCSV	−0.4	−1.02	0.01	355
Flunitrazepam	0.5 M KNO$_3$ (pH 3.9)	CPE	DPCSV	—	−0.69	0.4	356
5-Fluorouracil	Buffer solution (Na$_2$B$_2$O$_7$ + HClO$_4$, pH 7.8)	HMDE	DPASV	−0.17	+0.10	20.0	357
Lauril sulphate	1 M NaOH	HMDE	DPCSV	−0.90	−1.30	0.7	281
Lipamide	1 M NH$_4$Cl + 0.5 M NH$_4$OH + 2 10^{-5} M Co^{2+}	HMDE	d.c.SV	−0.60	−1.35	0.2	358
Methotrexate	0.05 M phosphatic buffer solution (pH 2.5)	HMDE	DPCSV	−0.10	−0.52	2.0	359
Methotrexate	0.1 M phosphatic buffer solution (pH 4)	CPE	d.c.SV	+0.40	+0.75	10.0	360
Mitomycin C	Phosphate buffer solution (pH 9) or mixture H$_3$BO$_3$ + NaOH (pH 10.2)	HMDE	DPCSV	−0.20	−0.18	2.0	361
Antibiotic ciprofloxacin	Britton-Robinson buffer (pH 4)	CPE	d.c.SV	Open circuit	0.95	20	362
Novobiocin	Boratic buffer solution (pH 11.6)	SMDE	DPCSV	−1.00	−1.38	1.2	354
Oxytetracycline	0.5 M NaClO$_4$ (pH 2)	HMDE	DPCSV	−0.50	−1.06	2.0	363
Paracetamol	Acetic buffer solution (pH 4.7) in 10% methanol	GCE	DPCSV (a.c.SV)	−0.10	−0.55	12.0	266
p-Phenylbenzophenone	Phosphate buffer solution (pH 6.8)	HMDE	CV	—	1.18	26.0	364
Streptomycin	0.01 M NaOH	SMDE	DPCSV	−1.20	−1.58	5.0	354
Thiocytozine	1 M NaClO$_4$ + HClO$_4$ (pH 2)	MFE	d.c.SV	−0.20	+0.12	50.0	365
6-Thioquonine	1 M H$_2$SO$_4$	RGCE	DPASV	+0.30	+0.83	100	362
Trypane blue	Britton-Robinson buffer solution (pH 2) or 0.1 M H$_3$PO$_4$	SMDE	DPCSV	−0.30	−0.92	5.0	366
Thyram	0.076 M NH$_4$OH	HMDE	DPCSV	−0.20	−0.84	0.03	367
Vitamin K	Acetic buffer solution (pH 4.2) in 5% methanol	HMDE	SWSV	−0.10	−0.26	1.0	368

chloride, and of benzene and dimethyl formamide (DMF); that of diethyldithiocarbaminate complexes of the same metals in the mixtures (1:1) of ethanol and benzene, toluene, and chloroform. Ammonium nitrate, chloride, rhodanide, or acetate was used as the electroconducting component. It was found that as the concentration of the organic ligand used to form extracted complexes of the metals studied was raised, the determination sensitivity of the SV method was usually lowered, a fact which is explained by kinetic limitations existing at the concentration stage due to the effect of the preceding chemical reactions.

To improve the detection limit in analysis of organic extracts, for example, in the determination of zinc, the "ammonium amalgam effect" was proposed: when an ammonium salt solution is electrolyzed, an ammonium amalgam is formed at a mercury dropping electrode at a sufficiently negative (-2.5 to -2.6 V) potential, with the amalgam decomposing rapidly forming a froth (ammonia and hydrogen). The electrode surface is enlarged, accompanied by an increase in the accumulation rate of the analyte element whose ions precipitate simultaneously with ammonium ions. When recording an anodic voltammogram, the mercury drop electrode is restored to its initial volume and, consequently, the concentration of the analyte element in the bulk of the electrode rises markedly, this enhancing sensitivity of the SV method.

In a series of their studies, Vydra and co-workers [242, 370] used a SEAM employing a glassy carbon rotating disk electrode to determine some metals in nonaqueous media that contained extracts of the metals.

Describing general principles of a hybrid version of SV or SCP extraction, Vydra [371] recommends the use of nonaqueous media for improving the detection limit through double concentration and for enhancing selectivity. However, the organic component of the extract can influence significantly the service characteristics of the electrodes used, especially of graphite and glassy carbon ones, leading finally to a dramatic degradation of the metrological characteristics (reproducibility and validity) of the analysis results, because, all other things being equal, the preelectrolysis time is usually longer than that in aqueous electrolytes when organic extracts are analyzed.

4.4. ANALYSIS OF NATURAL AND INDUSTRIAL MATERIALS

Figure 4.2 specifies principal applications of the SEAM in analysis of natural and industrial materials [372]. Such fields of application of the SEAM as clinical tests and analysis of medicines, food, biological materials, and, of course, environment (primarily natural waters and sewage) indicate the role of the SEAM in general trends of development of modern analytical chemistry.

In this section we shall consider specific features of the SEAM as applied

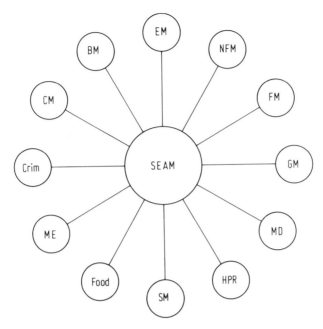

Figure 4.2. Major materials analyzed by SEAM: EM—environmental materials; NFM—nonferrous metallurgy materials; FM—ferrous metallurgy materials; GM—geological materials; MD—medicines; HPR—high-purity chemicals and reagents; SM—semiconductor materials; Food—foodstuffs; ME—microentities; Crim—criminalistics; CM—clinical materials; BM—biological materials.

to each of the specified subjects of analysis and determine the possibility of a general approach to the solution of analytical problems for each group of the materials analyzed.

As was noted above, SEAM have been developing for more than 30 years and such a great number of methodological papers have been written and such an enormous amount of information on the analysis of specific natural and industrial materials has been obtained over this period that to reflect it in all its entirely would require several large-volume monographs. Practical achievements of the SEAM were summarized at each stage of their development and every author tried to systematize the information on the use of the SEAM in the analytical practice accumulated by a given time. This was done in 1972 in a monograph [38]. The tradition was maintained in 1982 [14]. The papers [373, 374] deserve special attention, since they presents in tabular form almost all the then currently known methodological developments of the SEAM. The analytical quality of the SEAM is discussed in all of the monographs dedicated to this problem and published in the 1970s and 1980s. When preparing this monograph, the authors encountered a complicated problem of an ultimate reflection of the achievements and potentialities of the SEAM in analysis of various substances and materials, avoiding simultaneously unnecessary repetitions and concentration on minor problems.

We decided therefore to refrain from considering data that were generalized and discussed by other authors and refer the reader to the corresponding literature. We focussed mainly on the use of the SEAM in analysis of materials which have been so far inadequately covered in today's literature on the SEAM.

4.4.1. Analysis of Metallurgical Materials

Until recently, the SEAM was extensively applied to the analysis of materials of nonferrous metallurgy, although the number of papers on this subject decreased compared to the 1970s.

Most papers deal with analysis of nonferrous metals of different purity. As a result, we have presently a wide arsenal of techniques used to determine various microimpurities in practically all basic nonferrous metals. Besides, several ways of determining the same microimpurities by the SEAM have been developed for some matrices. Depending on the algorithm of particular techniques, pre-concentration method of the microimpurities to be determined, and type of the response registration, the lower detection limits attainable by the SEAM in the determination of trace concentrations vary from $n \times 10^{-8}$ to $n \times 10^{-4}\%$. In this respect, the SEAM are comparable, as applied to nonferrous metallurgy, to spectrophotometric and atomic absorption methods and have unquestionable points in their favor as regards determination of some elements (Bi, Sb, Cd, Sn, Pb). It should be borne in mind that in many cases the SEAM provide for simultaneous determination of several micro-impurities from a single voltammogram, for example, of Zn, In, Cd, Pb, and Cu in aluminum, Zn, Cd, Pb, and Bi in nickel, Te, Sb, Bi, and Pb in copper, while the vast majority of spectrophotometric and atomic absorption methods allow determination of only one element. Achievements in the analysis of some nonferrous metals are summarized in Table 4.7 [373].

It is natural to pose a question if it is possible, subject to such broad variations in the properties of the test matrices of nonferrous metals, to work out some generalized approach to the development of the SEA techniques for the trace analysis of the metals? The answer is negative from the viewpoint of optimization of conditions of some operations involved in determinations; however, some general approaches to the analysis of nonferrous metals by the SEAM do exist.

In the first place, the development of a SEA technique starts with the

Table 4.7. Analysis of Some Metallurgical Materials by SEAM [373][a]

Analyte	Variant	Supporting Electrolyte	E_{el}(V)	E_p(V)	Electrode	Material Analyzed	$c_{min} \times 10^4$ (%)
Ag	d.c.SV	0.1 M KNO$_3$ + 0.01 M EDTA	−0.3	+0.26	CPE	Copper and its alloys	5
Ag	d.c.SV	1 M KNO$_3$ (pH 1–2)	0.0	+0.38	IGE	Nickel	1
Ag, Cu [377]	d.c.SV	HF + HNO$_3$ (preelectrolysis)	−0.2 (Ag) −1.0 (Cu)		GCE	Tantalum	0.5 (Ag) 0.1 (Cu)
Ag [378]	d.c.SV	1 M Ga(NO$_3$)$_3$	0.0	+0.30	RGCE	Gallium	0.02
Ag [379]	a.c.SV	0.05 g L^{-1} CdCl$_2$	−0.65	+0.12	IGE	Cadmium	0.2
Ag	d.c.SV	0.1 M LiClO$_4$ in acetonitrile	−0.3	+0.18	SMDE	Mercury	2.0
As	d.c.SV	2 M HCl + 0.35 M Hg^{2+} + 0.4 M KI	−0.4	−0.7	Hg/CE	Iron ores	0.01
As	d.c.SV	3 M H$_2$SO$_4$	−0.7	−0.15	AuE	Silicon	—
As	a.c.SV	3 M HCl + 0.015 HI + 5 × 10^{-4} Cu^{2+}	−0.4	−0.72	SMDE	Gallium	0.03
As, Sb	d.c.SV	2 M H$_2$SO$_4$	−0.3 (As) −0.05 (Sb)	+0.21 +0.34	Au/GE	Copper	0.1
As	SCP	7 M HCl	−0.35	+0.04	Au/GE	Copper	0.5
As	SCP	7 M HCl + 10 M N$_2$H$_5^+$	−0.35	+0.04	Au/GE	Iron, nickel, chromium copper, tin, lead	
As	d.c.SV	4 M HCl	−0.5	+0.4	AuE	Tin	0.5
Au	d.c.SV	0.1 M HCl	−0.3	+0.45	CPE	Copper and copper products	0.1
Au	d.c.SV	0.5 M HClO$_4$ + 4 M HCl	+0.5	+1.16	CPE	Silicon	0.01
Au	d.c.SV	0.2 M KCl + 5 M HCl	−0.5	+0.38	IGE	Antimony, iron ores	0.01
Au	d.c.SV	1 M HCl	−0.4	+0.42	IGE	Iron, nickel, chromium tin	0.1
Bi, Sb	d.c.SV	1.1 M HClO$_4$ + 0.2 M TU + 2.1 M H$_2$SO$_4$		0.0 (Sb) +0.1 (Bi)	SMDE	Copper alloys	1.0
Bi, Pb	a.c.SV	6% NH$_4$Cl + 20% citric acid + 1 × 10^{-4} M Hg^{2+}	−0.9	−0.57 (Pb) −0.18 (Bi)	IGE	Ferroalloys, iron, niobium	0.1 (Pb) 0.02 (Bi)
Bi, Pb Zn	d.c.SV	1 M HCl(Bi) 0.5 M H$_2$SO$_4$ (Pb, Zn)	−0.3 (Bi) −1.2 (Zn, Pb)	−0.18 −0.92 −0.52	SMDE	Aluminum	0.1 (Bi, Pb) 0.5 (Zn)

91

Table 4.7. (*Continued*)

Analyte	Variant	Supporting Electrolyte	E_{el}(V)	E_p(V)	Electrode	Material Analyzed	$c_{min} \times 10^4$ (%)
Bi, Cu	d.c.SV	0.25 M HCl + nM NiCl$_2$	−0.45 (Cu) −0.30 (Bi)	−0.3 (Cu) −0.12 (Bi)		Nickel	
Bi, Sb, Pb	d.c.SV	1 M HCl + 5 × 10^{-4} M Hg^{2+}	−1.0	−0.55 (Pb) −0.28 (Sb) −0.15 (Bi)	IGE	Copper	0.1 (Pb) 0.03 (Sb, Bi)
Bi	d.c.SV	0.75 M HCl + 1 × 10^{-4} M Hg^{2+}	−0.8	−0.19	IGE	Antimony oxide	2
Bi, Sb	d.c.SV	0.2 M HCl (Te)	−0.45 (Te)	+0.37	IGE	Refined silver	0.05 (Bi, Sb)
Te [380]		0.2 M HCl + 1 × 10^{-4} M Hg^{2+} (Sb, Bi)	−0.5	−0.28 (Sb) −0.15 (Bi)			0.1 (Te)
Bi, Sb	a.c.SV	1 M HCl	−0.28	−0.22 (Sb) −0.15 (Bi)	SMDE	Cadmium and zinc	1.2 (Sb)
Bi, Cu [381]	d.c.SV	1 M HF + 1 M HNO$_3$ (preelectrolysis) HCl 5 M (electrodissolution)	−1.0	−0.24 (Cu) −0.17 (Bi)	GCE	Zirconium	0.01
Bi, Sb	d.c.SV	1 M HClO$_4$ + 0.1 M TU + 0.2 M ethylacetate		−0.1 (Sb) 0.0 (Bi)	SMDE	Tin	10 (Sb) 4 (Bi)
Cd, Pb, Cu, Zn	d.c.SV, DPSV	Na tartrate (pH 3.5)	−1.2	−0.98 (Zn) −0.65 (Cd) −0.50 (Pb) −0.05 (Cu)	SMDE, Hg/CE	Zirconium, zirconium oxide	2.0 (Zn) 0.05 (Cd)
Cd, In, Zn	d.c.SV	0.01 M KCl + ethylene-diamine	−1.6	−1.1 (Zn) −0.75 (Cd) −0.65 (In)	Hg/CE	Lead	0.1 (Zn) 0.01 (In, Cd)
Cd	d.c.SV	1 M KCl	−0.8	−0.7	SMDE	Zinc electrolytes	0.1
Cu, Pb	SCP	0.1 M KNO$_3$	−1.0	−0.52 (Pb) −0.12 (Cu)	SMDE	Cadmium	0.5

Element	Technique	Supporting electrolyte	E_1	E_2 (species)	Electrode	Sample	Concentration
Cu, Cd	d.c.SV	0.8 M HCl + 0.2 mmol l^{-1}	−1.0	−0.7(Cd), −0.25(Cu)	IGE	Zinc oxde	1
Cu, Sb	d.c.SV	6 M HCl	−1.1				
Cu, Pb, Cd [383]	d.c.SV	1 M KOH + 0.2 M ethylenediamine		−0.38(Cu), −0.74(Pb), −0.92(Cd)	SMDE	Mercury	0.2(Cd, Pb), 0.4(Cu)
Ce	d.c.SV	0.2 M HNO$_3$ + 0.08% phytin	−1.7		IGE	Lanthanum oxide	3.0
Fe	d.c.SV	2% citrate K + 2% KOH + 0.1 mmol L^{-1} Hg^{2+}		−0.75	IGE	Aluminum, silicon oxide	0.1
Mn	d.c.SV	0.2 M HMO$_3$ + 3 mmol L^{-1} KIO$_3$	+1.4	−0.9	IGE	Nickel	0.8
Mn	d.c.SV	1 M NaCl	−1.9	−1.12		Lead, mercury	0.01
Pb	SCP	0.1 M HClO$_4$ (pH 2)	−1.0	−0.52	RGCE	Thallium	0.4
Pb	d.c.SV	1 M HClO$_4$	−0.9	−0.50	CPE	Steels and iron	1.2
Pb, Cu	d.c.SV	1 M KCl (pH 3–4) + 0.02 mmol L^{-1}	−1.0	−0.5(Pb), −0.23(Cu)	GCE	Aluminum	0.25(Pb), 0.5(Cu)
Pb, Tl	d.c.SV	0.1 M K tartrate + 0.01 M EDTA (pH 4.7)	−0.8	−0.42	SMDE	Silicon	0.2(Pb)0.5(Tl)
Pt	d.c.SV	2 M HCl + 0.02 M EDTA + 0.03 M Sn^{4+}	+0.2	−0.36	CPE	Copper-nickel concentrates	0.02
Sb	d.c.SV	1 M HCl + 0.02 mmol L^{-1} Hg^{2+}	−0.5	−0.21	IGE	Zinc ores	2
Sb	d.c.SV	1 M HCl + 1 × 10^{-4} M Hg^{2+}	−0.45	−0.20	GCE	Nickel, ferromolybdenum, ferroniobium, ferrotungsten ferromanganese	2
Sb	a.c.SV	1 M HCl + 1 × 10^{-4} M Hg^{2+}	−0.55	−0.18	IGE	Copper electrolytes	1.0
Sb	d.c.SV	1 M HCl + 2 × 10^{-5} M Hg^{2+}	−0.35	−0.18	IGE	Ferroalloys	0.1
Sb	d.c.SV	1.2 M HCl + 1 × 10^{-4} M MG	+0.8	+0.6	IGE	Copper	0.08(Sb)
Sb, Sn	d.c.SV	2 M HCl	−0.5(Sb), −0.8(Sn)	−0.22(Sb), −0.56(Sn)	SMDE		0.03(Sn)
Sb	d.c.SV	7 M HCl	−0.5	−0.25	RGCE	Arsenic, bismuth, lead	0.05

Table 4.7. (*Continued*)

Analyte	Variant	Supporting Electrolyte	E_{el}(V)	E_p(V)	Electrode	Material Analyzed	$c_{min} \times 10^4$ (%)
Se, Te	d.c.SV	2.1 M NH₃ + 1.3 M HCl + 1 M NH₄Cl(Se) or 1 M NaOH(Te)	W/o external potential	-0.95(Se) -1.2(Te)	SMDE	Zinc concentrate	0.1
Se, TE	d.c.SV	0.2 M HNO₃ + 0.5 M NH₄NO₃	-0.1	$+0.64$(Te) $+0.90$(Se)	Au/GCE	Copper	0.1
Se, Te	d.c.SV	1 M HCl + Cu²⁺(Se) 1 M HCl	0.6(Se) -0.5(Te)	$+0.22$(Se) $+0.3$(Te)	IGE	Copper	0.01
Se	d.c.SV	M(NH₄)₂SO₄ + 0.4 M EDTA	-0.1	-0.81	SMDE	CdSiAs₂, ZnSiAs₂	0.08
Se	a.c.SV	1 M H₂SO₄ + 500 µg mL⁻¹ Cu²⁺ + 0.2 mL 5-% K₂Cr₂O₇	-0.4	-0.78	SMDE	Antimony, gallium	0.04
Tl	a.c.SV	1 M K tartrate + 0.05 mmol L⁻¹ Hg²⁺ (pH 8.6)	-0.9	-0.62	IGE	Nickel	0.5
Sn	a.c.SV	0.5 M H₂C₂O₄ + 0.1 mmol L⁻¹ MB	-0.75	-0.54	SMDE	Ferroalloys	0.5
Sn	d.c.SV	1 M HCl + 1·10⁻⁴ M Hg²⁺(1); acetate buffer (pH 7) + 0.6 M KCl (2)	-0.85	-0.62	IGE	Copper, zinc	0.1
Zn	d.c.SV	0.2 M acetate buffer	-1.45	-1.3	SMDE	Antimony chloride	0.03

[a]Potentials are given relative s.c.e.; TU—thiourea; MG—malachite green; MB—methylene blue.

94

estimation of the matrix composition which determines its electrochemical and chemical properties. Quite obviously, the more negative the redox potential of the matrix metal, the greater the possibility of direct determination of traces in the metal by the SEAM. Conversely, the more negative the redox potential of the analyte, the less the changes of determining the analyte microimpurity by the SEAM without preliminary separation of the micro-impurity and the matrix metal.

If we arrange the test metals in the ascending order of their standard redox potentials, we obtain the following series: Ba, Al, Ni, Zn, Cd, and so on, where the succeeding metals can be, in principle, determined in the matrix of the preceding metal without preliminary separation.

The general rule formulated has some limitations, because discharge-ionization potentials depend on the composition of the supporting electrolyte and can vary significantly due to complexation and hydration processes taking place in the solution, that is these potentials can become close or different for some analytes and ions of the base metal. This change in the discharge-ionization potentials of the components of the test solution often presents the analysts with an additional task of enhancing the resolving power of the SEAM. At the same time, the effect of discharge-ionization potential variation opens up the possibility, using widely complex-formation reactions, of not only improving the resolution of responses of the trace analytes but also of successful suppression of the matrix effects.

As the redox potential of the base metal shifts toward positive values, of the choice of a rational method of separating the traces and the matrix metal becomes important. For example, in the determination of Zn, In, and Cd in lead by the SV method, or of Cd, Pb, Bi, Se, and Te in copper, the method of preliminary separation of traces, which determines to a great extent the analytical characteristics of the technique, is of primary importance.

So, we face the problem concerning the choice of a rational method used to separate traces from the base metal, after all means of avoiding this operation, which is undesirable from the methodological and metrological standpoints, have been exhausted. The procedure of preliminary separation usually complicates the analysis technique, making it more time-consuming and laborious. Besides, introduction of additional operations in a technique is associated with an increase in the error of the results obtained by the technique, since every operation makes its contribution to the random component of the procedural error and adds to the probability of the systematic error occurring in the analysis results owing to uncontrolled losses of the analyte or pollution of the test sample with the analyte.

It follows from the foregoing that the number of operations aimed at preliminary separation should be as small as possible. This is the first and foremost rule to be adhered to when devising techniques of this type.

An example of these approaches is simultaneous determination of Zn, Cd, and In in lead, where the base metal is deposited at the anode as PbO_2 by electrolysis of the test sample solution and subsequently the traces are determined in the electrolyzed solution by the SV method [373]. A similar method of determining traces is exploited for determination of Cd and Pb in copper, where the base metal is deposited electrolytically at a platinum mesh electrode at a controlled potential and the traces are determined in the electrolyzed solution by the SV method using a preplated mercury graphite electrode. In these examples the general scheme of the technique is quite simple: dissolution of the sample—electrolysis—analysis of the electrolyzed solution by a SEAM. Here, a possible source of the systematic error is the interference effect of the residual amounts of the base metal on analytical responses of the microimpurities to be determined. In both examples, electrolysis lowers the content of the base metal in the solution to the negligible level. A systematic error arising from pollution of the samples of analytes can be eliminated if one uses reagents of appropriate purity or introduces a correction found from a blank test.

In many cases such a simple method of separation of microimpurities is hardly practicable in the analysis of nonferrous materials by the SEAM. And here we should note one more feature of the SEAM techniques.

Elimination of the matrix metal effect is undoubtedly the principal task in the development of the SEAM techniques, but in doing so account should be taken of possible mutual effects of the analyte and accompanying microimpurities deposited simultaneously. For example, in the determination of Pb and Bi in copper alloys, use is frequently made of codeposition of hydrated forms of the given microimpurities with $Fe(OH)_3$, $La(OH)_3$, $Th(OH)_4$, and so on in the ammonia medium (pH 11). In this case, the deposit contains, along with Pb and Bi, microquantities of Sn and Sb which give rise to a systematic error in the determination of Pb and Bi by the SEAM. A possible interference action of the components of the test sample on analytical responses of the analytes can be eliminated in several ways at different stages of the determination procedure. This can be achieved immediately at the stage of analysis of the solution obtained after separation of the matrix metal by varying the solution composition, that is, by setting up conditions removing the effect of the interfering component, which is achieved by the use of the complexation reactions described above.

Thus, to eliminate the influence of Sn and Sb on the analytical responses of Pb and Bi, in the example cited above the determination is performed against the background of 1 M NH_4Cl containing citric acid, in which Sn and Sb are electrochemically inactive. A 1-M KOH solution containing 0.25 M ethylene-diamine [373] can be used for this purpose.

Besides, at the stage of the deposit electrodissolution one can also use procedures directed at improving the resolving power of the method, such as change of the electrolyte or stop of the potential scan. The former procedure consists in the application of a supporting electrolyte which is optimal in composition for the analyte at the stage of ionization of the concentrate. This elimination of the interference effects is used extensively in the works by Kaplin.

The other procedure amounts to the stopping of the potential scan as soon as the maximum current of the element oxidizing at a more negative potential is attained. The potential scan is resumed after the ionization current of the component drops to the level of the residual current in the voltammogram. This procedure precludes superposition of the analytical responses of the concentrate components and close-valued ionization potentials (Fig 4.3). As seen from Fig. 4.3, when the potential scan is stopped, resolution of the analytical responses of the trace components enhances considerably and it becomes possible to eliminate a systematic error in the determination of In and Pb arising from mutual superposition of their peaks.

The interference effect of the microcomponents of the test samples on the analytical responses of the analytes can be eliminated also at the stage of separation of these elements from the matrix metal, that is, the separation procedure should include separation of unwanted components of the sample along with the base metal.

For example, determination of Zn, Cd, Pb, and Bi in nickel and copper–nickel alloys includes ion-exchange separation of these microimpurities through multiple-effect preliminary evaporation in the presence of hydrochloric acid and it is possible to remove almost completely trace Sn and Sb in the form of volatile chlorides from the sample solution.

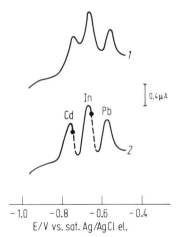

Figure 4.3. Voltammograms recorded without stopping the potential scan (*1*) and with the potential scan stopped (*2*) at potentials marked with dots (the dashed line is registered during the potential scan stop). $\tau_{el} = 1$ min; $v = 0.4$ V/min; $c_{Cd} = c_{Pb} = 5 \times 10^{-7}$ M; $c_{In} = 2 \times 10^{-6}$ M. Supporting electrolyte; 1 M KCl; Hg/CE.

In many cases the separation method is selected so as, conversely, to extract the maximum information about the microimpurities present in the test metal, that is, to effect group trace deposition. In the majority of the examples given above the techniques rested precisely on the principle of simultaneous determination of a maximum possible number of elements from a single solution. We should admit however that techniques aimed at the determination of one element by the SEAM prevail in nonferrous metallurgy. These techniques, first, can employ a broader range of procedures of separation and concentration of the microimpurity to be determined and, second, facilitate optimization of the determination conditions of one component compared to that of several elements. The choice of the group or individual deposition of microimpurities in the SEAM, similar to other analysis methods, is governed by specific requirements but, in our opinion, when it is necessary to determine several microimpurities and this can be done either by simultaneous or successive analysis of the solution obtained, preference should be given to the methods of group deposition.

Nonferrous metallurgy is one of the branches where the SEAM proved to be a reliable and highly sensitive means of analytical control.

For a long time the SEAM have not been used in analysis of materials of ferrous metallurgy, because in ferrous materials the content of elements usually answers the capacity of the d.c. or a.c. polarography ($n \times 10^{-3} - n \times 10^{-2}\%$). The first study dealing with the determination of lead in superalloys by SV with mercury preplated GE [375] showed that the SEAM opened up the possibility of determining traces in many materials of ferrous metallurgy, omitting preliminary separation from the major alloy components.

The technique [375] made it possible to curtail the time of a single analysis from 6 h to 45 min, excluding all operations associated with separation of trace lead. Such a result could be expected since the major component in materials of ferrous metallurgy is iron. In a reducing medium, iron does not impede determination of Pb, Sn, Bi, Cu, and Sb, since it becomes Fe^{2+}. Other major components of ferroalloys are either sufficiently electronegative metals (Ni, Co, Nb, Cr) or electrochemically inactive over a wide interval of the working potentials of mercury and solid working electrodes (Zr, Ta, B, etc.).

A unified techniques has been developed for SV determination of Pb and Bi in binary alloys of Fe, Ni, and Co with Nb [376]. It has been found that in the above alloys Pb and Bi can be determined when $c_1 = 2 \times 10^{-4}$ and $5 \times 10^{-5}\%$, respectively. No significant effect on the results of Pb and Bi determination was noted at whatever the content of Ni^{2+}, Co^{2+}, and Fe^{3+} or at 0.2% Cu. The interference of copper with the analytical response of bismuth is eliminated if the potential scan is stopped as soon as the maximum dissolution current of copper is attained (-0.25 V).

This technique has been extended to cover, practically without any

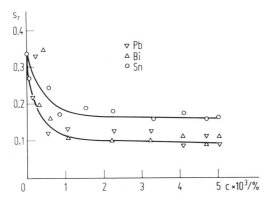

Figure 4.4. Relative standard deviation as a function of the Pb, Bi, and Sn concentration in ferroalloys when determining impurities by Hg/CE SV.

changes, binary ferroalloys, for example, ferrochrome and ferroboron as well as ferromolybdenum and ferrotungsten.

When Pb and Bi are determined by this technique, the relations $s_r = f(c_{Pb}, c_{Bi})$ are given, irrespective of the composition of the test alloys (except FeW and FeMo), by a single curve (Fig. 4.4) over the interval 1×10^{-4}–$5 \times 10^{-3}\%$ of the analytes.

A method used to concentrate antimony as low-soluble compounds with triphenylmethane dyes on the surface of a graphite electrode also offers the possibility of developing a unified technique for analysis of ferro- and super-alloys, excluding the stage of preliminary separation of antimony from other components of test samples. Data on the application of the SEAM in analysis of ferrous materials are presented in Table 4.7.

4.4.2. Analysis of Microentities

Film coatings and crystals (see Fig. 4.5) are often used in electronic and semiconductor engineering. Properties of these materials and of the instruments and devices manufactured on their basis depend largely on the content and distribution of dopants (in the formula of the test material, these are shown on the right of the matrix and are separated with a dot) and impurities in individual layers and microvolumes. Layers, surface areas, and microvolumes of crystals and films belong to microentities whose analysis presents a problem, since one has to deal with extremely small samples of test materials $(10^{-3}$–10^{-5} g) and, correspondingly, with a limited volume of the sample solution. A change from analysis of macroentities to analysis of microentities, which is associated with a considerable (4–5 orders of magnitude) decrease

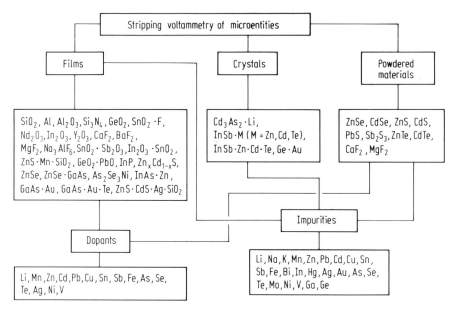

Figure 4.5. Stripping voltammetry of film coatings and crystals

in the mass of the test substance, unfailingly leads to worsening of the detection limits. Simultaneously with a drastic decrease in the mass of test samples, a problem of raising the number of analytes determinable at a time from a single sample presents itself, that is, a problem of transition to this or that form of multicomponent analysis. Requirements imposed on accuracy and reproducibility change greatly, too. Thus, we have a specific and essentially new field of analysis, where a certain part can be taken by the SEAM thanks to their rather low detection limit, the possibility of varying widely the determination conditions, and a sufficiently large number of determinable elements. The problem of using the SEAM, in particular SV, for analysis of microentities was solved successfully in the studies by Kaplin and co-workers [374, 383] who developed and realized in practice this novel analytical branch. A great number of analysis procedures has been worked out for microentities (Table 4.8).

On the basis of the specific features of the microentities analyzed and a great number of microelements to be determined in these, the authors [383] worked out the fundamentals of a scheme of a multielement successive stripping electroanalysis that combines the use of a single initial sample and optimum conditions of the final determination of individual elements or of their groups that are close in electrochemical properties. Examples of SV use are listed in Table 4.8.

Table 4.8. Analysis of Microentities by SV

Ord. No.	Entity Analyzed (weight, g)	Analytes	Determination Interval, $\Delta c \times 10^{-6}$ (%)
	One-Substance Films		
1.	Li_2O (10^{-2}–10^{-3}) 374	Zn, Cd, Cu, Pb, Bi	1–10
2.	CaF_2 (or BaF_2, MgF_2) ($n \times 10^{-4}$) [383]	V, Ag	0.5–8.0
3.	CaF_2 (BaF_2, MgF_2)	Zn, Pb, Cd, Sn, Cu, Sb, Bi	1–5
4.	CaF_2 (BaF_2, MgF_2) [383]	Ni	1–5
5.	ZnS (ZnSe, CdS, CdSe) (10^{-2}–10^{-3}) [374]	Ni	1–5
6.	$Zn_xCd_{1-x}S$ on GaAs substrate (10^{-3}–10^{-4}) [374]	Ga, As	0.5–50
7.	Y_2O_3 [374]	Cu, Pb, Cd	1–10
8.	Nd_2O_3 (10^{-2}–10^{-3}) [374]	Cu, Pb, Cd	1–10
9.	Si (SiO_2) on InAs substrate (10^{-3}–10^{-4}) [384]	In, As	0.4–10
10.	SiO_2 (Si_3N_4) on Si substrate (10^{-3}–10^{-4}) [374, 383]	Au, Zn, Cd, Ag, Sn, Pb Ga, Sb, Bi, Te, Mn, Fe	1–10
11.	Geo ($n \times 10^{-3}$) [374]	Zn, Cd	1–10
12.	HfO_2 (ZrO_2, TiO_2) ($n \times 10^{-3}$) [383]	Zn, Cd, Pb, Sn, Cu, Sb	0.1–5
13.	SnO_2 (Sb_2O_3) (10^{-3}–10^{-4}) [374]	Zn, Cd, Pb, Cu, Sb	0.5–15
14.	As_2Se_3Ni ($n \times 10^{-3}$) [383]	Ni	5–12
15.	$VGdS_3$ ($n \times 10^{-3}$) [385]	V	0.5–10
	Alloyed Films		
16.	GaAs·Au (10^{-3}–10^{-4}) [383]	Au	0.3–1.2
17.	GaAs·Au·Te (10^{-3}–10^{-4}) [383]	Au, Te	0.3–1.8
18.	GaAs·Zn (10^{-3}–10^{-4}) [383]	Zn	10^4–10^3
19.	GaAs·Cd (10^{-3}–10^{-4}) [383]	Cd	10^4–10^3
20.	GaAs·Se (10^{-3}–10^{-4}) [272]		10^4–10
21.	GaAs·Te (10^{-2}–10^{-4}) [386]	Te	10^2–10
22.	GaAs·Fe ($n \times 10^{-3}$) [387]	Fe	10^2–10
23.	In_2O_3·SnO_2 ($n \times 10^{-3}$) [374]	Zn, Cd, Pb, Cu	1–10
24.	Yn_2O_3·SnO_2 [374]	In, Sn	10^3–10^2
25.	SnO_2·Sb_2O_3 ($n \times 10^{-3}$) [374]	Sn, Sb	10^3–10^2
26.	$SnO_2Sb_2O_3$·ZnO ($n \times 10^{-3}$) [374]	Sn, Sb, Zn	10^3–10^2
27.	SnO_2 (Mn·ZnS) ($n \times 10^{-3}$) [388]	Sn, Mn, Zn	10^{-3}–10^2
28.	SnO_2 (Mn·ZnS) [388]	Cd, Pb	1–10

Table 4.8. (*Continued*)

Ord. No.	Entity Analyzed (weight, g)	Analytes	Determination Interval, $\Delta c \times 10^{-6}$ (%)
	Multilayer Films		
29.	Al/V/Glass ceramics $(n \times 10^{-4})$ [385]	V	1–10
30.	SiO$_2$/V/Glass ceramics $(n \times 10^{-4})$ [383]	V	1–20
	Crystal Microlayers		
31.	ZnCd·Te	Zn, Cd, Te	1–10
	InSb·Zn	In, Sb, Zn	
	InSb·Cd	In, Sb, Cd	
	InSb·Te	In, Sb, Te	
	$(n \times 10^{-3})$ [374]		
32.	Cd$_3$As$_2$·Li $(n \times 10^{-3})$ [383]	Li	80–100
33.	Microsamples Nb$_2$O$_5$ (Ta$_2$O$_5$) $(10^{-2}–10^{-3})$ [374]	Bi, Cu, Pb	1–50

The technique of successive multielement analysis of microentities supposes changing of working electrodes and supporting electrolytes for the determination of individual elements or of their groups, as well as selective separation of elements.

It is known from practical applications of the SEAM (see Tables 4.1 and 4.2) that the principal supporting electrolytes used for determination of traces of heavy metals are HCl, KCl, NH$_4$F, and NH$_4$Cl solutions of different concentration. A 0.1-M KCl background can be used for determination of Mn, Cd, Zn, Pb, and Cu by the SV method employing a mercury film electrode; metals such as Sn, Sb, and Bi are determined better in HCl; 1–2 M HCl and a graphite indicator electrode are recommended for determination of Au and Te. When devising a technique of multielement analysis, it is particularly important to take into account the mutual influence of the elements in the concentrates on graphite or mercury electrodes over the determinable concentration interval $10^{-7}–10^{-8}$ g/mL. An investigation of the Au–Te system showed that these elements could be determined simultaneously over the interval specified above only when $c_{Te} : c_{Au} \leqslant 1 : 20$. At other concentration ratios intermediate peaks are registered and determination of both elements by the SV method is systematically underestimated. The interference effect of gold ($c_{Au} = 10^{-7}–10^{-8}$ g/mL) can be eliminated by evaporation of the test solution to dryness with reduction of Au^{3+} to metallic gold.

To remove the basic components when GaAs is analyzed, it is recommended that gallium be extracted from a 6-M HCl medium using diethyl ether. Other elements (Mn, Zn, Cd, Pb, Sn, Cu, Bi, Sb, Te) either cannot be extracted or are extracted in insignificant amounts (1–1.5 rel.% in the case of Zn). Gold is extracted simultaneously with gallium. Arsenic should be removed, like gallium, from the solution, because its reduction leads to formation of an As film on the electrode surface which increases the residual current. To achieve an almost complete removal of arsenic, the GaAs film is dissolved in a mixture (3:1) of HCl and HNO_3, As^{5+} are reduced to As^{3+}, and the sample solution is additionally evaporated twice with HCl. A successive multielement stripping electroanalysis of the GaAs film can be represented schematically as follows:

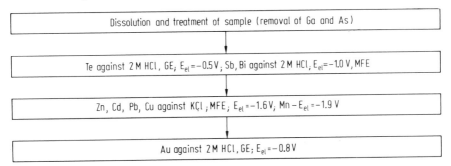

Similar schemes of the successive multielement SV analysis were proposed for SiO_2, Si_3N_4, and Si films. Practical application of such schemes allows one to estimate the content of impurities and dopants both in the entire volume of the film and in the surface layers of films and crystals.

Comparing potentialities of the SEAM (specifically SV) and other methods of analysis used to check composition of microentities, the following conclusions can be drawn.

So far, SV of microentities is inferior to mass spectrometry and the neutron-activation method by an order of 1–2, as regards detection limit of the most microcomponents and by a factor of 2–3 as regards the number of determinable elements. However, with present requirements for high-purity materials and substances, films and crystals produced on their basis, the possibility of determining the absolute contents of 10^{-9}–10^{-11} g is sufficient when actual microcomponents are analyzed.

The need to determine 50–60 elements during a "review" analysis of microentities is a relatively rare case. More frequently it is necessary to determine 10–15 elements in a single microsample. These are usually Li, K, Na, Mn, Zn, Cd, Pb, Cu, and Fe and a small number of other elements (As, Se, Te, Au, Pd, Ga) in materials of microelectronics and quantum electronics; "dyeing" impurities (Fe, Co, Ni, Mo, W, V, Ag, Cu, Mn, Cr) in luminophors, optoelectronic materials, fibrous light conductors, and so on. At the given stage, such microentities can be analyzed by SV. As is known, the SV equipment and techniques are much simpler and more readily available than equipment and techniques used in mass spectrometry and neutron-activation analysis. Note also the comparatively short time required for analysis of microentities by the SV method. Thus, subsequent to removal of the Si layer and extraction of the matrix element, determination of, for example, 10^{-8}–10^{-9} g of arsenic in silicon requires 0.5–1 h; in the case of the neutron-activation analysis, keeping a silicon sample after irradiation alone accounts for 3–5 days.

Comparing the spectrochemical analysis and SV of microentities, let us emphasize a higher absolute sensitivity of the latter and the possibility of performing a quantitative analysis without the mandatory use of solid reference samples. The problem of the multicomponent analysis of microentities has already been considered.

A definite advantage of SV over the atomic adsorption method in determination of microentities is the possibility of determining up to 10–12 elements from a single microsample using the scheme of multielement successive analysis.

4.4.3.   Analysis of Environmental and Biological Materials

Beginning in, the 1970s, the importance of the SEAM rose sharply owing to an acute necessity of monitoring the environment, biological materials, and food, and the need for clinical tests. Similar requirements are imposed on analytical methods when these materials are analyzed: low detection limits of toxic elements; multielement analysis; the need to determine not only the total content of toxic elements but also to obtain qualitative an quantitative information about physicochemical forms (including degree of oxidation) of the analyte (analysis aimed at the determination of various physicochemical forms of elements is called "speciation analysis," a term coined by Nürnberg [389]); the possibility of automatic continuous checking (monitoring) and distant control of the analysis procedure.

We shall not touch upon medico-social problems but concentrate on the items which are related directly to the SEAM as one of the promising ways of solving problems associated with monitoring of environmental pollution and analysis of biological and clinical materials. The largest group of the environmental materials, for which the SEAM are especially efficient, is represented by natural waters and sewage.

4.4.3.1.   Analysis of Waters

The principal problem of checking the composition of waters is connected with complexity and time variation of the composition of waters to be analyzed. On the face of it, natural waters are solutions of mineral salts of various concentrations. In reality, such a solution contains a great variety of diverse inorganic and organic compounds, including ionic, molecular, and surface-active ones, solid and colloidal particles capable of adsorbing trace analytes, and finally systems of microorganisms and biologically active compounds which can incorporate the microelements to be determined. For the same reason, the trace analysis of natural waters is a complicated task whose solution calls for special techniques and a metrological approach.

Up to the present time we do not know the predominating physicochemical form of many heavy metals in which these occur in natural waters. However, toxicity and biological acitivity of an element depend on its particular physicochemical form (the latter is shown in Table 4.9) rather than on its concentration.

It is customary to assume [390, 391] that free (hydrated) ions of heavy metals are toxic. Metal complexes or metals bound with colloidal particles are less toxic, although, for example, soluble lipid complexes of cadmium or copper are as toxic as free ions of metals owing to their ability to rapidly penetrate into a biological medium with subsequent dissociation and release of a free metal ion.

A change in the form of trace elements can alter toxicity greatly. The majority of investigations dedicated to the toxic action of heavy metals on fish and other aquatic organisms show that the free form of metal ions is indeed the most poisonous and toxicity is due just to the concentration of this form of the metal and not to its total concentration [390, 391]. Toxicity of metals usually drops with increasing density and salinity of waters mainly due to the formation of complexes of metal ions and inorganic ligands. Pure sea and river waters generally contain extremely low concentrations of free metal ions. The majority of dissolved metals are present in the form of

Table 4.9. Some Physicochemical Forms of Metals in Waters [391]

Form	Example
Hydrated metal ions	$Be(H_2O)_4^{2+}$
Inorganic complexes	$Pb(H_2O)_4Cl_2$
Low-soluble salts	PbS, $ZnCO_3$
Stable organic complexes (salts)	Cu-glycinate
	Cu-fulvate
Ions (Cu^{2+}, Pb^{2+}) adsorbed on inorganic colloids (Fe_2O_3, MnO_2)	
Organic colloids (humic acid)	
Mixed colloids (humic acid-Fe_2O_3)	

nontoxic complexes (e.g., with fulvic acids) or are adsorbed on colloidal particles (e.g., on Fe_2O_3 particles coated with humic acids). Besides, the so-called detoxication reactions take place in natural waters—transforming free ions of metals into their nontoxic forms.

A critical deterrent to the use of the SEAM in the analysis of waters is adsorption of organic substances on the surface of indicator electrodes [390, 391]. Adsorbates form a layer which can affect the diffusion rate of the ions to be determined and, sometimes, suppress the diffusion flows and alter the kinetics of the electrode process, thus resulting in a nonlinear relationship between the ionization current and the preelectrolysis time or concentration. The processes of adsorption–desorption of organic dipoles on the surface of a mercury electrode may be the source, with the SV method, of "tensammetric" peaks, especially when high-frequency voltammetric methods (a.c. and pulse variants) are used [390, 391]. These peaks do not have the faradaic component but can be easily mistaken in the SV method for peaks of metal ionization currents, because in the analysis of natural waters such peaks can appear at the same potentials as the analytical responses of Cd, Pb, or Cu [391]. However, tensammetric peaks can be easily differentiated from the metal ionization peaks if one changes to the d.c. conditions, where such effects are not observed. Besides, tensammetric peaks vanish after the destruction of organic substances.

A detailed study has been made [392] on the influence of organic components on the determination by the derivative SV method of Cd, Pb, and Cu in solutions simulating natural waters. Three parameters were considered as sources of information: maximum of the anodic current time derivative (response h); response h as a function of the metal concentration (sensitivity S); and the concentration (c) found by the standard additions method.

Table 4.10 presents the abovementioned three parameters referred to (in %) the respective parameters of a reference solution that does not contain the organic substances studied. As is seen, all the organic substances (except glucose) affect the response and sensitivity of lead and copper determination. Here, an essential factor is the time elapsed between introduction of the metal ions to be determined into the solution of the organic substance and the moment of measurement.

The effects observed can be explained by the complex-formation processes proceeding slowly in the bulk of the solution. Thus, the influence of the composition of the model solutions containing organic compounds on the determination of the metal ion concentration by the SV method should be studied some time after the analyte ion is brought in contact with the organic component in water.

For fulvic and humic acids and polyethylene glycol, a decrease in response and sensitivity is accompanied by a systematic underestimation of the concen-

Table 4.10. Effect of Some Organic Compounds on Determination of Lead and Copper in Model Solutions by LSSV[a]

Organic Compound	Lead						Copper					
	After Addition to Solution			After 6 h			After Addition to Solution			After 6 h		
	h/h_o	S/S_o	c/c_o	h/h_o	S/S_o	c/c_o	h/h_o	S/S_o	c/c_o	h/h_o	S/S_o	c/c_o
Fulvic acid	−56	−44	−20	−11	+78	−50	+30	+30	const	−20	−15	−7
Humic acid	−22	−2	−20	−44	+39	−60	−10	−5	−7	−40	−36	−7
Poly(ethyleneglycol)	+33	+21	+10	−33	+11	−40	−30	−34	+7	−30	−30	const
Urea	const	const	const	−11	−1	−10	+20	+19	const	+10	+3	+7
Petroleum products	−22	−22	const	−22	−22	const	−20	−30	+13	−20	−30	+13
Sodium dodecylsulfonate	const	const	const	+33	+33	const	−40	−48	+13	−10	−5	−7
Phenol	−22	−29	+10	+33	+21	+10	−10	−16	+7	−10	−16	+7
Diethylamine	−11	−19	+10	−22	−29	+10	const	+6	−7	const	−6	+7
Glucose	const	const	const	const	const	const	const	const	const	const	const	const

[a]Supporting electrolyte: 0.1 M KNO_3; $v = 0.4$ V/s; $\tau_d = 1$ min $c_{Pb} = 0.5\ \mu$g L^{-1}; $c_{Cu} = 1.5\ \mu$g L^{-1} [392].
h—response; S—sensitivity; c—concentration determined; − decrease; + increase; const—unchanged.

tration (see Table 4.10) determined by the additions method. In other cases, the concentration of lead and copper found by the additions method does not practically show systematic deviations.

The results obtained in [392] on the influence of organic substances on the accuracy with which concentration of heavy metal ions is determined in water using SEAM (SV in the case under consideration) suggest that the change of the analytical response cannot be considered as the criterion solely of the complex-formation processes taking place in natural waters. The most comprehensive information about these processes occurring in solutions and on the electrode surface can be obtained taking into account three parameters (response, sensitivity, and concentration found). This approach makes it possible to detect the formation of stable electroinactive complexes in water and to distinguish between these processes and inhibition of electrode reactions on the electrode due to adsorption.

SEAM can be used in combination with other methods to assess the ratio of active (labile) forms of metals and their electroinactive (inert) forms in water samples, as well as to determine these forms.

Florence [391] notes that the SV method is used most widely to assess the ratio labile form/inert form (L/In) for Cu, Pb, Cd, Zn, Mn, Cr, Tl, Sb, and Bi. The so-called pseudocolloids of heavy metals, that is, colloidal particles of Fe_2O_3, MnO_2, humic acids, and so on, that adsorb ions of heavy metals can be considered as a special type of metal complexes capable of contributing to the determination of labile forms.

The speciation analysis includes also determination of the degree of oxidation of the element in the solution. The degree of oxidation is a parameter of utmost importance, since it often defines toxicity and adsorption behavior of microimpurities in water. For example, toxicity of arsenic(III) is 1000 times higher than that of arsenic(V), while toxicity of thallium(III) is 10 times lower than that of thallium(I). The SV method has been used to determine the degree of oxidation of $Fe^{2+,3+}$, $Tl^{1+,3+}$, $Sn^{2+,4+}$, $Sb^{3+,5+}$, $As^{3+,5+}$, $Se^{4+,6+}$, and $Mn^{2+,4+}$ in solutions.

For some elements, determination of the degree of oxidation of the form present in the solution is a special case of the determination of the L/In ratio, because one of the metal forms may be electroinactive and the other may be electroactive over a given range of the indicator electrode potentials. For example, As^{5+}, Cr^{3+}, and Sb^{5+} are electroinactive under certain conditions. The content of the labile form of an element in the presence of the given inert forms can be assessed quantitatively using a common SV variant. The total content of the element is determined after an appropriate chemical treatment of the sample and the content of the element with the inert degree of oxidation is found from the difference in the estimates of the total content and labile forms of the elements.

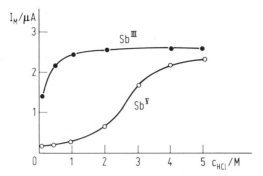

Figure 4.6. Maximum oxidation currents of Sb^{III} and Sb^V as a function of HCl concentration [391].

Some metal ions are electrochemically inert because of an extremely high level of hydrolysis in the most media; for example, Sb^{5+} becomes labile on strong acidification of the sample [391]. Figure 4.6 shows the behavior of Sb^{3+} and Sb^{5+} depending on the acidity of the medium when the SV method is used. The content of Sb^{3+} can be determined in 0.2 M HCl, while the total concentration of antimony can be found in 6–8 M HCl, and then the content of Sb^{5+} is easily determined from the difference in the results of the preceding two determinations.

One of the major limitations of the speciation analysis with employment of the SEAM is the impossibility of estimating the concentration of individual forms of ions. For example, it is impossible to distinguish between the labile forms of cadmium (Cd^{2+}, CdCl$^+$, or CdCl$_3^-$) which can coexist in river water. The registered stripping peak of the current will be due to the presence of all of the abovementioned forms. This "common" behavior of responses is also typical of other types of speciation analysis, for example for ion-exchange chromatography, dialysis, or ultrafiltration [390, 391].

The information about the forms in which elements are present in water can be derived both by immediate testing of samples using the SEAM (e.g., when the L/In ratio is determined) and after various operations of sample pretreatment. In this case the SEAM are used simply as highly sensitive methods of analysis.

A most complete scheme of the speciation analysis is given in [389, 391]. The scheme includes filtration of samples through a membrane filter with pores 0.45 um in diameter, separation of microelements using the Helex-100 resin. UV irradiation of samples, and SV determination of the concentration.

Using this scheme, the SV method can be exploited to determine qualitatively concentrations of metal ions in various forms: free ions, labile organic and inorganic complexes, labile forms adsorbed on organic and inorganic colloidal particles, inert organic and inorganic complexes, and inert forms adsorbed on organic and inorganic colloids.

The authors [70] put forward a scheme for the determination of the physicochemical forms of Zn, Cd, Pb, and Cu in river water and sewage. The scheme is based on different lability of the forms, which are subdivided into extremely labile, moderately labile, weakly labile, and inert. Use is made of filtration, ion exchange on the Helex-100 resin, and acidic decomposition of organic components in the sample in combination with the SV method. It is shown that the water samples tested contain Cd and Zn as extremely and moderately labile forms, Cu as moderately and weakly labile forms, and 20–70% Pb is present in a weakly labile form.

The development of the SEAM-based speciation analysis led to appearance of new variants of this group of methods. Thus, for example, a method was proposed, which was called pseudopolarography, where the form of the trace metals in solutions is determined from the relationship between the maximum currents of the metal anodic dissolution and the preelectrolysis potentials (see Fig. 4.1). The method has been applied to determination of the forms of cadmium and lead in sea water. It is shown that when pH ⩽ 5, the chloride complex of lead prevails in sea water. The same method was exploited to examine the influence of complexation on with copper(II) reduction in sea water. It is inferred that at pH of water in natural conditions copper(II) is reduced with kinetic difficulties caused by the formation of hydroxide, carbonate, and dihydrocarbonate complexes.

A modified variant of the SV method has been proposed. It is based on the reduction of metal on the electrode surface at different potentials, subsequent shift of the potential to the region of the limiting diffusion reduction current at a rate of 500 mV/s, and recording of the anodic curve at a rate of 50 mV/s. The source of information is the relationship between the maximum anodic current and the preelectrolysis potential (neopolarograms). The method has been used to study the interaction of Cd^{2+} and Pb^{2+} with chloride-ions and to determine stability constants of the chloride complexes of these metals [390, 391].

The complexation capacity of organic substances of natural waters is given in [389–391]. Water samples were titrated with a metal salt solution and the amount of metal not bound into complexes was determined by the SV method. It is recommended to use copper salts for titration, because copper forms most stable complexes with the majority of organic ligands. Results obtained from the study of the interaction between Cd, Cu, and Zn ions and fulvic and humic acids in sea water by the differential pulse SV method show that at the natural pH the complex forms of metals change with time in the sample taken [390, 391].

Complexation varies depending on the place of sampling, depth, and biological efficiency of the region. A decrease in the complexation capacity of the sample relative metal ions was observed [390] during the first 6 h after sampling for waters featuring high biological efficiency. Sea-water fulvic acids interact with the metal ions studied differently: interaction with cadmium is nearly absent, while that with Zn^{2+} is significant; interaction with copper depends on the fulvic acid structure. The mechanism of interaction between metal ions and humic acids is complicated and is not clearly understood.

A study was made [393] of the influence of cationic, anionic, and neutral surfactants and humic acids on the parameters and potentials of responses of Cu, Pb, and Cd when a special solution simulating sea water is analyzed by the SMDE SV method. When the concentration of surfactants is less than or equal to 0.1 ug/L, their influence on metal responses is insignificant. With higher concentrations, of surfactants, responses decrease as a rule, but sometimes increase for Cu and Pb. When the concentration of humic acids in the solution exceeds 1%, responses of Cu and Cd decrease and the response of Pb remains practically unchanged. The authors explain these data by adsorption of surfactants and humic acids on the electrode surface and complexation of the metal ions in the solution.

The authors [345] present their findings on the forms of mercury in polluted river waters. A titration method in combination with the DPSV was used to estimate the complexation capacity of water. The titration curves exhibit three straight portions, each corresponding, as the authors believe, to binding of mercury ions with some ligands. It was shown that even when pH < 3 mercury ions (70% of the total content) are bound into strong complexes with organic ligands (fulvo- or aminocarbonic acids).

One of the most interesting results obtained from the speciation analysis of waters by the SV method is the detection of a correlation between the labile (under the SV conditions) forms of elements and their toxicity. However, the correlation between the content of the labile forms of copper and its toxicity found by the authors agrees only in part with the biotesting of waters. Florence [391] found that results of the SV determination of the labile forms of copper in sea water made at low negative potentials of preelectrolysis correlated with the concentration of the toxic forms of copper in specially prepared solutions close in composition to the test samples and containing natural complex-forming agents such as fulvic, humic, tannic, and other acids. When synthesized ligands (2,9-dimethyl-1,10-phenanthroline or ethyl xanthate) were introduced into solutions, the correlation was disturbed.

Let us sum up applications of SV in the speciation analysis of waters.

1. Among the analytical methods used to check the composition of waters, the SV method is one of the most powerful as regards analytical parameters

and informativeness. Considering the known mechanisms by which traces of elements in various forms penetrate into living organisms through biomembranes, it is possible to select the solution composition and electrochemical parameters of the SV processes such that the rate of metal deposited at the stage of preelectrolysis corresponds to the rate at which the trace metal is adsorbed by the biosystem.

2. The relation between the physicochemical forms of trace elements and their toxicity in water can be determined using the SEAM (primarily SV). This field of the speciation analysis application evidently lacks more refined schemes that would allow a more complete determination of the possible forms of trace metals in waters.

As was previously noted, after samples of natural water have been treated using some operations, under certain conditions, SEAM loses speciation-analysis properties and becomes a common highly sensitive technique of checking waters for the total content of trace metals.

The interference of organic substances occurring in the analysis of water samples by the SV method can be eliminated through destruction of the substances using UV irradiation [390, 391, 394], ozonation [395], γ-irradiation, and chemical ashing [396]. With UV irradiation, one should take into account possible losses of electropositive elements, including mercury, due to their reduction, Nevertheless, this method of removing organic components from waters is rather efficient. It makes it possible to improve accuracy of the water analysis results, to lower the detection limit achieved with the SV method, and to perform determinations without preliminary deaeration of solutions in some cases [394].

The authors [241] put forward an electrochemical method of preparing natural waters for SV analysis, which eliminates the interference of surfactants and complex-forming organic substances. The method is based on anodic oxidation of organic substances in the presence of chloride-ions.

The efficiency of the electrochemical pretreatment of waters can be judged from the derivative stripping voltammograms of Cu, Pb, and Cd. Electrochemical mineralization of samples is considered to be sufficiently efficient if the values of response and sensitivity (slope of the calibration curve) obtained after electrochemical pretreatment of the solution coincide with the values of the corresponding quantities obtained for simulated solutions which do not contain organic components.

To optimize the electrochemical pretreatment conditions, a study was made of the influence of the electrode potential, solution composition, and electrolysis time on the character of stripping voltammograms.

Figure 4.7, a displays responses of Cu, Cd, and Pb as a function of the electrode potential during anodic treatment of samples. Organic substances

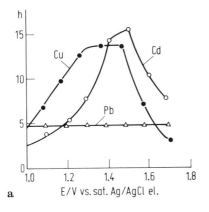

a

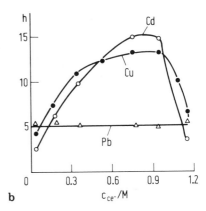

b

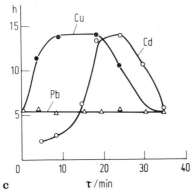

c

Figure 4.7. Responses of Cd, Pb, and Cu (in relative units) versus electrolysis potential at the oxidation stage (*a*), concentration of chloride-ions (*b*), and anodic oxidation time (*c*); $a - \tau_a = 20$ min; $c^0_{Cl^-} = 0.75$ M; $b - E_a = +1.4$ V; $\tau_a = 20$ min; $c - E_a = +1.4$ V; $c^0_{Cl^-} = 0.75$ M. SV conditions are: $E_{el} = -1.1$ V (Ag/AgCl$_{el}$) $\tau_{el} = 1$ min; $v = 0.4$ V/s; 1×10^{-4} M Hg (NO$_3$)$_2$ is added to the supporting electrolyte.

are destructed best over the interval 1.4–1.5 V. At this stage, 0.75–1 M (Fig. 4.7*b*) should be taken as the optimum concentration interval of chloride-ions. If the upper limit of the given interval is exceeded, responses of the elements decrease owing to the accumulation of chlorine in the solution, a circumstance which affects the SV determination of the elements studied.

As the time of the anodic treatment of water samples is increased from 10 to 25 min, ionization currents of Cd, Pb, and Cu rise in the stripping voltammograms (Fig. 4.7*c*), resulting from the destruction of organic complexes of the metals being studied in water. A further increase in the treatment time again leads to an accumulation of chlorine with the aftermath specified above.

Anodic oxidation of organic compounds represents an efficient treatment of waters both with low and high contents of organic substances. The use of a glassy carbon crucible which serves during anodic treatment of water samples both as the anode and the electrolyzer made it possible to enlarge substantially the working electrode surface and, consequently, to raise the electrolysis efficiency.

The SEAM that are most frequently used for analysis of waters are direct current (d.c. SV) and differential pulse (DPSV) voltammetry.

A derivative variant of the d.c. SV method employing recording of analytical signals at a high rate of the electrode potential scan was used to determine traces of Cu, Pb, Cd, Zn, Co, Mn, Cr, Te, As, and other elements in water [390]

An important advantage of the SEAM is the fact that here the presence of the salt background in waters is a favorable factor, while for AAS, for example, this is a limitation. Consider also the relatively simple control and low price of the SEAM equipment, as well as ease of instrument calibration and control over the validity of the SEAM analysis results.

For these reasons, in the last five years the SEAM have become the most widespread methods used to check various kinds of water: from sea and river waters to sewage and snow (Table 4.11).

4.4.3.2.   Analysis of Air, Soil, Plants, and Food

Analysis of air, aerosols, airborne dust, and the solid components of atmospheric precipitation (rain, snow) by various SEAM usually includes sufficiently long sampling operations connected with pumping of large volumes of air through absorption solutions for an air check and through filters for analysis of solid particles. Therefore the SEAM are unlikely to compete with sophisticated quick-acting gas analyzers. However, some successful applications of the SEAM in checking atmospheric air for such toxic metals as Cd, Hg, Cu, and Pb have been described.

For example, during international investigations of the background levels of pollution of the atmosphere, the authors [410] carried out a detailed chemical analysis of solid particles occurring in the air over a region of the Pacific Ocean far removed from industrial centers. The content of 15 elements was checked, out of which Cu, Pb, Cd, and Zn were traditionally determined by the Hg/CE DPSV. The results obtained by the SV method agree with those of AAS. The use of the SEAM (SV and SCP) for checking the composition of solid particles in atmospheric precipitation is envisaged by the normative documents in force in Germany for the determination of toxic Cd, Tl, Pb, Cu, Zn [411].

The possibilities of SEAM are so far insufficiently appreciated for the analysis of air (e.g., the methods can be used to advantage for the determination of As, Se, or Te in the air of working areas). At the same time, the approach to the development of the air analysis techniques on the basis of the SEAM has been essentially unified, a fact which simplifies the task. Considering the

Table 4.11. Analysis of Waters by SEAM

Place of Sampling (Water Type)	Analytes	Seam Variant	Working Electrode	c_{min} ($\mu g\,L^{-1}$)	Ref.
Indian Ocean	Hg	d.c.SV	HMDE	0.01	397
(deep layers, littoral waters)	Pb, Tl	SCP	Hg/CE	0.20	398
Seas					
Australian	Pb, Cd	DPASV	HMDE, Hg/CE	0.015 (Pb) 0.007 (Cd)	391
Baltic (near Sweden)	Zn, Cd, Pb, Cu, Bi	PSA	Hg/CE	0.006 (Bi) 0.01 (Cd, Pb) 0.10 (Zn, Cu)	399
Barents	Cd, Pb, Cu, Ni	DPASV	Hg/CE GCE(Ni)	0.05 (Cd, Pb) 0.20 (Cu, Ni)	400
Canadian	Cd, Pb, Cu	d.c.SV	Hg/CE	0.02 (Cd) 0.04 (Pb) 0.10 (Cu)	391
Norwegian	Cd, Pb, Cu	d.c.SV	HMDE	0.10 (Cd, Pb) 0.40 (Cu)	401
North (near England)	Zn, Cd, Pb	d.c.SV	Hg/AuE	0.05 (Zn) 0.01 (Cd) 0.05 (Pb)	391
Mediterranean	Zn	d.c.SV	HMDE	0.30	115
(near Italy)	Zn, Cd,	d.c.SV	HMDE,	0.02 (Cd)	391
(near Yugoslavia)	Pb, Cu		Hg/CE	0.05 (Pb) 0.12 (Zn, Cu)	
Black (near Ukraine)	Cd, Pb, Cu	d.c.SV (derivative)	Hg/CE	0.05 (Cd, Pb) 0.15 (Cu)	390
Japanese (near Japan)	Zn, Cu, Pb, Cd	DPASV	HMDE	0.20 (Zn) 0.04 (Cu) 0.01 (Pb, Cd)	402
Gulf of Mexico	Zn, Cd, Cu	DPASV	HMDE, Hg/CE	0.04 (Cd) 0.20 (Zn, Cu)	391
San Diego Bay	Zn, Cu	d.c.SV	Hg/CE	0.60 (Zn) 0.10 (Cu)	397
Estuaries of					
Guanabara (Brazil)	Pb	d.c.SV	HMDE	0.025	391
Trent (England)	Zn, Cd, Pb, Cu	a.c.SV	HMDE	0.20 (Cd) 1.0 (Zn, Pb, Cu)	391
Huelva (Spain)	Zn, Cd, Pb, Cu	PSA	HMDE	0.30 (Zn) 0.02 (Cd) 0.10 (Pb) 0.15 (Cu)	403
Drinking water					
Germany (Ruhr)	Zn, Cd, Pb, Cu	DPASV	HMDE	0.20 (Cd) 0.50 (Pb) 2.0 (Cu) 5.0 (Zn)	404
(München)	Cu	a.c.SV	CPE	0.50	243
India (Bombay)	As	d.c.SV	HMDE	1.0	405

Table 4.11. (*Continued*)

Place of Sampling (Water Type)	Analytes	Seam Variant	Working Electrode	c_{min} (μg L^{-1})	Ref.
North Korea	Zn, Cd, Pb, Cu	PSA	HMDE	2.0 (Cd, Pb) 5.0 (Zn, Cu)	406
Russia (Ural)	Pb, Cu, As, Se	d.c.SV	Hg/CE	0.2 (Pb) 0.5 (Cu) 0.04 (As, Se)	390
USA (Chicago)	Cd, Pb, Cu, Hg	DPASV	Hg/CE	0.05 (Cd, Pb, Cu) 0.002 (Hg)	407
Japan (Tokyo)	Sb	DPASV	HMDE	0.004	391
(Cobe)	Cd	DPASV	Hg/CE	0.005	391
Rain water					
Germany (Ruhr)	Zn, Cd, Pb, Cu, Tl, Ni, Co	DPASV	HMDE	—	389
Belgium (Brussels)	Zn, Cd, Pb, Cu, Mn, Co, Ni, Se	DPASV, PSA	HMDE	—	391
Tap water					
Germany (Ruhr)	Zn, Cd, Pb, Cu, Tl	PSA	HMDE	0.70 (Zn)	389
Poland	Cu, SCN$^-$	d.c.SV	HMDE	0.02	391
Artesian wells (USA)	Tl	DPASV	Hg/CE	0.05	408
Greenland snows	Cd, Pb, Cu, Ni	DPASV	Hg/CE GCE (Ni)	0.10 (Cu) 0.02 (Cd, Pb) 0.05 (Ni)	391
Waste water					
Great Britain	Zn, Cd, Pb, Cu, Cr, Ni	DPASV	Hg/CE GCE (Ni, Cr)	50 (for all metals	409
Russia	As	d.c.SV	AuE	0.05	390

latest advances in SEAM as regards determination of organic compounds, in the immediate future one might expect the appearance of techniques concerned with the determination of toxic organic substances in the air and atmospheric aerosols.

General problems and prospects for the use of the veltammetric methods of analysis, including SEAM, for checking the composition of soil, plants, agricultural products, and food are considered detail in the review [412]. The number of papers dedicated to the use of SEAM for analysis of soil is small (see Table 4.12). These methods have not yet received such wide recognition in this field as in the analysis of waters. As with the analysis of air, the cause of such a situation is assumed to be the long time required to separate the analyte components from the matrix. The separation operations involved in the SEAM-based techniques are usually connected with "wet" ashing or burning of samples.

Taking determination of trace thallium and lead in soil as an example, the authors [420] consider the possibility of a rational combination of DPSV and the most efficient methods used to concentrate the microimpurities to be separated. Ten standard samples of soil with a certified thallium content were analyzed. Subsequent to ashing of the sample, thallium was preconcentrated using (1) extraction of $TlBr_4^-$-with organic solvents, and (2) ion-exchange chromatography. It was found that the first method of concentration was good for acid and mildly alkaline soils as well as clay deposits with the thallium content not exceeding $2 \times 10^{-4}\%$. The second method is advantageous in the analysis of sandy, alkaline, and strongly alkaline soils. The techniques that involve ion-exchange chromatography are of interest because the relative error incurred in the determination of thallium by SMDE DPSV is independent of the matrix composition and varies between 3 and 22% when c_{Tl} is 700–50 ng/g.

To check the content of Cd, Cu, Pb, Cl^-, Br^-, and I^- in some soils (red soil, fluvial deposits), it was proposed [421] that SMDE a.c. SV be used. To improve the selectivity of determination, voltammograms should be registered with the second a.c. harmonic or in the high-freqency mode. The most accurate results were obtained for 400–600 ng/g of lead and copper in soil. This technique is extensively used in some Europian countries and the United States.

The Hg/CE SCP (PSA) methods were exploited to determine Cd, Pb, and Tl in air-dry samples of soil and ash [420]. Concentrate the metals to be determined on a rotating glassy carbon electrode plated with mercury in situ −0.95 V. When thallium is determined, preelectrolysis is carried out in the presence of EDTA. The developed technique was used for certification analysis of standard samples of ash.

Traces of Zn, Cd, Pb, Cu, Te, and Mn in plants were determined by SMDE a.c. SV [422]. To check the validity of the analysis results, use was made of standard samples of leaves of fruit trees. The content of traces determined by the SV method agrees well with the certified content. A comparison of the SV data with the results obtained by the flameless atomic-absorption method showed that the SV method is characterized by a higher accuracy (relative error ± 12 and $\pm 21\%$, respectively) but requires careful removal of the organic matrix. Otherwise, the SV method systematically yields results which are too low.

To determine traces of Zn, Cd, Pb, and Cu rice, apples, and other foodstuffs, the MFE SWSV method was used [423].

Table 4.12. Determination of Organic Compounds and Drugs in Environmental and Biological Materials by HMDE AdSV[a]

Material Analyzed	Analyte	Concentration Range (g L^{-1})	Separation Method	Ref.
Natural waters	Dibuthyl tin chloride	5×10^{-5}–3×10^{-6}	–	349
	N,N'-Dimethyl-4-amino-3-methylazobenzene	5×10^{-7}–1×10^{-5}	TCL or extr.	351
	Caprolactam	4×10^{-5}–2×10^{-3}	TLC	343
Natural and sea waters	Petroleum components	3×10^{-5}–1×10^{-3}	–	413
	Purines (guonine, hypoxantine, adenine, xantine)	1×10^{-5}–8×10^{-4}	HPLC	337
Natural waters and soil	Thiram	8×10^{-6}–4×10^{-4}	extr.	367
Natural water and snow	Folic acid	5×10^{-6}–1×10^{-4}	extr.	355
	Chlordiazepoxide	8×10^{-7}–3×10^{-5}	CC	
Blood plasma and serum	Amethoptirine	4×10^{-6}–6×10^{-4}	CC or FIA	340
	Benzodiazepin derivatives (diazepam, oxcazepam, fluorazepam, clonazepam, bromazepam, clolazepam)	6×10^{-6}–5×10^{-4}	FIA	414
	Daunorubicin	6×10^{-6}–8×10^{-4}	extr.	348
	Thiocytosine	6×10^{-8}–1×10^{-6}	sorp.	365
	Clothiazepam	1×10^{-5}–2×10^{-3}	CC	415
	Triazolam			
	Cyadox	1×10^{-6}–5×10^{-4}	FIA	416
	Streptomycin and related antibiotics (aritromycin, novobiocyn)	5×10^{-7}–2×10^{-5}	–	354

Urine and blood			
Antifilhorzial antimony compounds (antimoline, astibone, bilchorcide, stibonfene)	2×10^{-5}–1×10^{-3}	CC	417
Methothrexate	2×10^{-7}–1.2×10^{-3}	extr.	359
Mitomycin C	2×10^{-7}–1.6×10^{-5}	–	361
Lipoic acid, lipoamide	1×10^{-7}–6×10^{-5}	CC	358
5-Fluorouracil	2×10^{-5}–1×10^{-4}	extr.	357
Marcellomin, tethrocoine	5×10^{-7}–2×10^{-6}	HPLC	418
Vitamin K_1	1×10^{-9}–1×10^{-6}	CC	368
Paracetamol	1×10^{-4}–7×10^{-3}	FIA	266
2-Thiobarbituric acid (2-TBA): 5-ethyl-5'-(1-methyl-buthyl-2-TBA; 1-methyl-5-ethyl-5'-(1-methyl-propyl)-2-TBA, chlor-hydrate 2-methyl-TBA	10^{-5}–6×10^{-3}	CC	419
Thiocarbosamides (TCA): 5,6,7,8-tetrahydro-3-methyl-quinoline-8-TCA; 5,5,7,8-tetrahydro-3-methyl-quinoline-8-(N-dimethyl)-TCA	4×10^{-5}–5×10^{-3}	CC	416

[a]TLC—thin layer chromatography; extr.—extraction; HPLC—high performance liquid chromatography; CC—column chromatography; ECC—extraction column chromatography; FIA—flow injection analysis; sorp.—sorption.

The possibility is shown to determine 0.5 ng/mL–6 µg/mL of tin in some foodstuffs using the SMDE SV method [424]. The authors considered potentialities of some methods in preliminary separation and concentration of low contents of tin (ion-exchange chromatography, extraction with tropomine in toluence, deposition as SnO_2) depending on the test matrix composition.

As is seen from this data, the SEAM are used in the analysis of soil, plants, and food mainly to determine traces of toxic heavy metals; when determining the edibility of foodstuffs, it is extremely important to know the content of microimpurities of toxic organic substances. Some progress in this direction has been achieved through the use of adsorptive concentration [281] (see, e.g., Tables 4.6 and 4.12).

4.4.3.3. Analysis of Biological Materials

In recent years the analysis of biological materials has become, along with analysis of water, one of the main fields of SEAM application, a fact which is favored to a certain extent by a common approach to the solution of analytical problems in the given fields.

Biological examinations, including analysis, can be performed using two methods: in vivo (i.e., directly in a living organism), for example, by monitoring continuously the composition of a biological liquid of blood plasma in an organism without sampling, or in vitro (i.e., with sampling of the test material).

With the SEAM, the in vivo approach presents some difficulties associated with the introduction of the electrode system, which is intended to register analytical responses, directly into living tissue, with provision of conditions for the concentration of the analyte component on the electrode surface. Conversely, the in vitro method in combination with the SEAM is highly promising, since these methods provide for analysis of microvolumes of biological and clinical materials, avoiding destruction of the biomatrix. Thus, the papers cited in [374] describe the successful application of the Hg/CE or SMDE SV method for the determination of microcontents (5×10^{-6}–5×10^{-7} g/L) of heavy metals (Cu, Pb, Zn) in blood samples of 20 to 100 µl. The techniques proposed do not differ essentially from those suggested for analysis of natural waters (see Section 4.4.3.1). This was clearly demonstrated by Batley and Farrar [425] who determined Cu and Pb in sea water and blood under the same conditions by the SMDE DPSV method. The technique developed by those authors is characterized by the values $c_{min} \approx 1$ µg/L (Cu) and 0.6 (Pb) ug/L.

The development of the flow-injection analysis applicable to microvolumes of solutions offered new possibilities of the analysis of biological liquids.

An automatic flow-injection SV blood-testing system capable of performing

1000 blood tests a day was designed and clinically tested [426]. The system allows determination of Cd, Pb, and Cu in blood at a level ≈ 3 ug/L. The indicator electrode is a disk electrode made of pyrolytic graphite and plated with mercury in situ. When the system operates under optimum conditions, the time of a single analysis is 34 s. The relative standard deviation is ± 3 and $\pm 7\%$ (rel.) for 200 and 5 ug/L respectively.

Some biological and clinical materials cannot be checked by the SEAM without destruction of the organic matrix; then it is necessary to choose rational schemes of analysis including, as a rule, the stage of sample ashing.

Investigations carried out in some countries on the principal characteristics of the techniques used to determine trace Hg, Cd, Pb, Cu, Zn, Fe, Mn, and Ni in algae, zooplankton and fish meal and also in bottom sediments [427] made it possible to recommened AAS or SV methods for the determination of Zn and Cu, and the SV method, as the most accurate and sensitive, for Cd and Pb.

The number of elements determined by the SEAM in biological materials has increased recently owing to the newly developed techniques of determining Bi, Sb, As, and Se (see Table 4.12). In the United States, extensive use has been made of a technique of direct determination of trace selenium(IV) in blood, which consists of dissolving the sample in the Lumatom solvent (quarternary ammonium base in an organic solvent) and subsequent determination by SMDE CSV. The technique has been recommended for the analysis of biological materials with a low content of As^{3+}, Cu^{2+}, Fe^{3+}, Pb^{2+}, and Zn^{2+}. The high accuracy of the analysis results obtained by this technique was confirmed by analysis of standard biosamples, for which no statistically significant discrepancies between the certified content of selenium and the content determined using the SV method were found.

In the last 3–4 years the SV method has gained popularity as a method to determine traces of toxic organic substance, drugs and their metabolites, pesticides, and so on in biological materials and clinical analysis [428–431].

The potentialities for SV in this sphere of analysis are illustrated in Table 4.12. The overwhelming majority of techniques involve HMDE DPSV. Given below are some of the most interesting examples. Thus, the possibility is shown of using AdSV to check biological processes taking place in biosystems. In [432], this method was exploited to study the interaction of immunoglobulin E and antiimmunoglobulin E in blood. Both compounds lend themselves to adsorptive concentration on a HMDE at $+0.1$ V (sat.cal.el.). The optimum conditions are assumed to be a 0.05-M phosphatic buffer solution (pH 7.4), $\tau = 5$ min, and $v = 10$ mV/s. Under these conditions, the calibration graph of immunoglobulin E is linear over the interval $1 \times 10^{-9}–1 \times 10^{-7}$ mol L^{-1}. The value $\tau = 3$ min is recommended for antiimmunoglobulin, with the other conditions being the same. The calibration graph of this compound is linear over an appreciably narrower interval of concentrations (1×10^{-9}–1.5×10^{-8} mol L^{-1}). Adsorbate reduction peaks (approx. -0.55 V) are registered under the DPSV mode.

The HMDE AdSV method was used to study the interaction between immunoglobulin and benzodiazepin in blood plasma [433]. The authors refined the determination conditions and recommended as the supporting electrolyte the one used in [432], $\tau = 3$ min, pulse amplitude 50 mV, and $v = 10$ mV s^{-1}. The detection limit of immunoglobulin A was 3×10^{-10}–7×10^{-9} mol L^{-1}.

The electrochemical behavior of bovine serum albumin was studied using cyclic voltammetry and the DPSV method in order to check the processes that take place in the blood of animals [434]. With a phosphatic buffer solution (pH 7.4–8.0) containing 1×10^{-8} mol L^{-1} of the abovementioned compound after adsorptive accumulation ($E = +0.15$ V, $\tau_{\alpha} = 50$–120 s) under the DPSV mode, two peaks were registered at -0.34 and -0.55 V, whose magnitude increases as the concentration of the compound is increased.

The authors [413] developed a technique for determination of cyadox and its metabolites (monoxide cyadox, desoxycyadox, quinoxaline-2-carbonylglycine) in animal blood plasma after separation of the compounds by column chromatography or extraction. The technique was used to monitor the transformations of cyadox and its excretion from the organism.

The AdSV method has been applied for the determination of anticoplastic agent methotrexate in blood plasma and tissues [359]. The compound is accumulated on a HMDE at -0.1 V (sat.cal.el.) against a 0.05-M phosphatic buffer solution (pH 2.5) and is determined in the DPSV mode, recording three peaks, out of which for analytical purposes the most pronounced and largest peak at -0.51 V, which is due to reduction of the pyrazine ring, is recommended. The detection limit is 2×10^{-9} mol L^{-1}, corresponding to 9 ng of methotrexate in 10 mL solution.

The number of studies made with the use of solid electrodes in this field is rather limited. In [352], the disk CPE DPSV method made it possible to determine microconcentrations of dopamine in urine samples, while the GPE DPSV was used successfully to determine 1×10^{-6}–1×10^{-8} mol L^{-1} of chlorpromazin in urine. Rotating GCE DPASV was exploited to determine 2×10^{-5}–1×10^{-3} mol L^{-1} G-mercarptopurine and G-thiogyanine in urine [362]. It can be seen from Table 4.12 that in the analysis of biological materials and clinical analysis the AdSV often serves as a highly sensitive means of detection in various types of chromatography (HPLC, TLC, CC) and flow-injection analysis. This branch of flow-through SEAM has been refined and advanced annually.

PHASE ANALYSIS OF SOLIDS

Solids are extremely complicated subjects of investigation and analysis. They frequently represent a conglomerate of crystalline and amorphous phases of constant and variable composition, contain nuclei (clusters) of different phases and lattice defects. More or less complete information about these entities can be extracted only if use is made of different methods. Let us consider one of the approaches employed in the analysis of a solid, namely, the electrochemical phase analysis which makes it possible not only to identify and to determine the amount of a particular phase but in some cases to determine structural defects of the entity analyzed. The importance of such information can hardly be overestimated, since the phase composition and structure defects of solids determine their technological properties, such as sintering, catalytic activity, and conductivity.

As distinct from the electrochemical phase analysis, where the actual electrochemical operations are, as a rule, only separation of phases and transfer of the required components into solution which is subsequently analyzed by any appropriate method, in SEAM the operations of separation and determination overlap in time and space. In the last 10–15 years, noticeable advances have been made in the study of the phase composition, degree of oxidation of elements, adsorbability, catalytic activity, and defect structure of solids using SEA methods [435–438].

The use of a carbon paste electrode opens up interesting possibilities. One such possibility was realized as early as the late 1950s [439–41], when a prototype of this electrode was probably used for the first time in a study of electroreduction reactions of aromatic compounds. A relationship between the cathodic potential that is set up during current flow and the nature and position of the substitutes in a solid compound molecule was established.

This technique was further developed by Songina and Barikov [109]. With this electrode, the solid studied is mixed with carbon powder and is introduced directly into the electrode reaction zone, thus providing for recording of electrical signals due to electrochemical transformations of the solid or, as will be shown in Chapter 6, to adsorption–desorption of the electrolyte decomposition products on the solid's surface. The authors called their electrode a mineral carbon-paste electrode, but subsequently the term

carbon-paste electroactive electrode (CPEE) was suggested [21], since potentialities of this electrode go beyond the study of minerals.

So far, there is not a single term to identify the method using CPEE. It
is referred to as voltammetry with a mineral carbon-paste electrode [109],
paste electrode voltammetry or bulk stripping voltammetry [14], solid-phase
or contact voltammetry [442], and stripping voltammetry of solids [21]. In
our view, the most appropriate terms are stripping electroanalytical chemistry
of solids (SEACS) or electrochemical analysis of solid-phase entities. At
present, SEACS includes two sections: phase analysis (Chapter 5) and examination of specific features of the structure of solids (Chapter 6).

These chapters treat the processes that take place at the solid—aqueous
electrolyte solution interface:

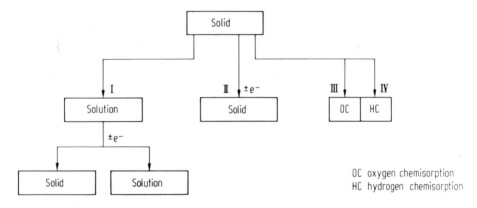

In all cases, hydration of the surface layer is a more or less pronounced
intermediate stage. Two limiting situations are possible: destruction of the
initial structure of the compound and its relative preservation. The data on
the initial structure we are interested it can be obtained, naturally, in the
latter case only. Signals of typical electrochemical transformations are formed
in the processes "solid–solution–solid" (I) and "solid–solid" (II). Process I
furnishes information mainly about the phase composition and some rather
limited information about structural peculiarities depending on the relation
between the dissolution rate and the compound structure. The kinetics of
process II is a strong function of the type of crystalline and electronic structure
of the compound and therefore the corresponding features are reflected in
signals of typical electrochemical transformations.

Electrochemical transformations of oxygen and hydrogen sorbed on the
surface of a solid during processes III and IV bear information essentially
about the electronic structure of the solid and the nature of its disorder.
These processes are considered in Chapter 6.

5.1. ALLOYS AND METAL-CONTAINING COMPOSITIONS

The pioneering paper describing the use of CPEE in electrochemical analysis was dedicated to a study of metal powders [109]; however, the number of papers concerned with this problem is relatively small. At the same time, materials of powder metallurgy represent excellent subjects of investigation and analysis with the use of CPEE, considering that these materials are highly dispersed and homogeneous, that is, no sample pretreatment is required.

A study was made of the electrochemical behavior of molybdenum and tungsten [443], rhenium [444], palladium [445], iron [446], arsenic [447], copper [448], and alloys Cd_4Sb_3, CdSb, ZnSb, Zn_4Sb_3, InSb, [449], and NiZn [450]. The kinetic parameters of electrooxidation of metals (Ag, Co, Ni) were investigated [14].

Extensive use is currently made of metal-graphite compositions which are produced by evaporation and subsequent condensation of one (or several) metal(s) on graphite particles in vacuum furnaces. Depending on the substrate temperature, the metal coating thus formed can be found in three different states (droplet, islet, and mixed). Hence, the method used to check the quality of a metallized material should be sensitive to the coating structure. From this viewpoint, the most promising is the method of CPEE stripping voltammetry [451] (Fig. 5.1). Although the mean content of copper in the samples is the same, the current–voltage curves are a different shape.

Figure 5.2 displays calibration curves for different types of copper coating on graphite. Sensitivity is at a maximum for islet coating and at a minimum for droplet coating.

The procedure of analysis of copper–graphite compositions by the CPEE voltammetry method consists of the following. The CPEE paste (e.g., 0.05 g test sample + 0.25 g uncoated carbon + 0.3 mL dibutylphthalate) is prepared. The paste is mixed dry and after addition of the hydrophobic liquid the anodic curve is registered from -0.4 V with the potential scanned linearly [451].

A highly sensitive CPEE SV method has been developed [452] for qualitative and quantitative phase analysis of nickel–aluminum and copper–aluminum-based Reney multimetallic skeleton catalysts. The method is used to determine Al, Ni, Fe, Cr, Ti, Cu, Zn, Cd, Sn, Nb, Pd, and intermetallic compounds $NiAl_3$, Ni_2Al_2, Ni_3Al, $FeAl_3$, Fe_3Al_5, $FeAl_2$, FeAl, Fe_3Al, CuAl, and Cu_2Al. Sensitivity to aluminum and zinc is much higher than to other metals.

The authors propose increasing the resolution of the method by stopping the potential scan after the oxidation current maximum is attained [452]. The anodic current of more electronegative phase is thereby lowered and

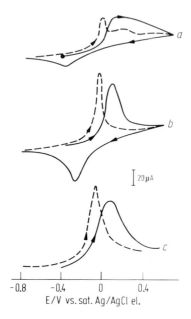

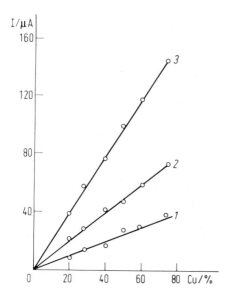

Figure 5.1. Anodic–cathodic–anodic (*a*, *b*) and anodic (*c*) voltammograms of electrochemical transformations of the droplet (*a*) and islet (*b*, *c*) copper coatings of particles of natural graphite powder [451]. Conditions: 0.5 M H_2SO_4; $v = 80$ mV/s; $E = -0.60$ V; (*c*) $\tau = 3$ min. Content of copper in the samples is 51 mass%.

Figure 5.2. Maximum anodic current versus copper content in metallized graphite powder for different types of coating [451] (0.5 M H_2SO_4; $v = 80$ mV/s): *1*, droplet coating; *2*, mixed coating; *3*, islet coating.

the electropositive phase becomes much more pronounced. The maximum anodic current is directly proportional to the concentration of the metal or intermetallic compounds in the paste and hence can be used in the quantitative phase analysis of metallic catalysts.

Qualitative and quantitative phase analyses of the Reney catalysts of various compositions are based on the identification of the oxidation peaks of the catalyst and those of the reference sample.

When CPEE is used for phase analysis of metals and alloys, it is expendient to use electrodes with a small surface ($0.03 \, \text{cm}^2$), because oxidation currents of metal powders are sufficiently high. This electrode provides for high selectivity of the method; curves of individual phases fall within a potential interval which is narrower than that of large-surface electrode.

An electroanalytical technique for the determination of precipitated phases in steels and alloys, the etching quantitative phase analysis method, has been used [453] for the investigation of cast samples.

In this method, the matrix must be in the passive state and the phase to be determined must be in the active state. The content of the phase is determined using the electric signal that arises during selective anodic dissolution (etching) of the phase. There are two requirements for the achievement of the selective anodic dissolution: (1) The difference in the composition of the separated phases must be large enough and (2) the polarization conditions must be suitable. In halide ion-containing electrolytes, the matrix is usually etched and the phases are usually extracted. But in basic solutions the matrix is usually in the passive state and the other phases are usually in the active state. When the polarization conditions (electrolyte, potential, temperature, etc.) are controlled, the dissolution rate and order of the phase dissolution are different, and a selective dissolution can be achieved.

An etching quantitative phase analysis method for the determination of M_6C and MC separately in high-speed steel has been developed [454]. A current peak arising over the potential interval of about $-200 \, \text{mV}$ (s.c.e.) in a 10% NaOH: 2% tartaric acid solution and caused by the dissolution of the M_6C phase may be used to determine the content of the M_6C phase. The MC, $M_{23}C_6$, and M_7C_3 phases do not interfere with the determination of M_6C. The current peak emerging over the potential interval of about $+1400 \, \text{mV}$ (s.c.e.) in 1% $H_2Cr_2O_7$ solution and caused by the dissolution of the MC phase can be used to determine the content of the MC phase. The M_6C, $M_{23}C_6$, and M_7C_3 phases do not impede the determination of MC. A simple analysis method without standard has been studied. The precision of the method is similar to that of the usual extraction method.

The analysis of tin-bismuth alloys using a new stripping technique, abrasive stripping voltammetry, has been described. Small amounts of the alloy are transferred by mechanical abrasion from the solid alloy onto the surface of a paraffin-impregnated graphite electrode. These traces are anodically stripped off using differential pulse voltammetry. The ratio of the tin and bismuth peak currents provide a measure of their relative content in the alloy. The determinations have standard deviations of about 5%. Peak potentials are reproducible within a 20-mV range [455].

5.2. CHALCOGENIDES

For analysis of chalcogenides, particularly of sufide materials, use is made of electrolysis and voltammetry employing an electrode consisting of the test substance, or voltammetry employing paste electrodes containing the test electroactive substance.

Sufide-anode electrolysis is used for separate extraction of components. In the case of the CPEE SV, the stages of separation and determination are combined. Sulfides are anodically oxidized as a rule to elementary sulfur and the corresponding cations go into solution. The electrochemical behavior of sulfides depends on their nature (Gibbs energy, crystal lattice energy, chemical interaction with electrolyte solutions). The greater the crystal lattice energy, the more positive potential is required for the compound oxidation. As regards the decrease of stability and oxidation potential, metal sulfides arrange themselves in the series $MoS_2 > ReS_2 > FeS_2 > CuFeS_2 > CuS > Cu_2S$, providing for the principal possibility of their determination when the compounds occur simultaneously.

Compounds of $Cu_{2-\delta}S$ composition with discrete intervals of the values of δ (0.05–0.09, 0.14–0.20, 0.32–0.35, 0.60–0.64) give a spectrum of cathodic and anodic signals [456] (paste electrode with conductive binder $2 M H_2SO_4$ is used). An analysis of the data allows one to calculate the values of δ for the series of nonstoichiometric copper sulfides given above, as well as to compute normal potentials of the redox systems and the Gibbs energy of formation of nonstoichiometric sulfides from chalcosine Cu_2S.

A study of the electrochemical transformations of copper(II) sulfide introduced into CPEE in a $4–10 M NH_3 + 1.5 M (NH_4)_2SO_4$ solution shows that oxidation is accompanied by formation of Cu^{2+} and SO_4^{2-} ions as well as polysulfides and sulfur.

The internal standard method [457] makes it possible to simplify considerably the analysis procedure and to extract a more complete information about the entitity studied. Figure 5.3 presents voltammograms of CuS, Cu_2S (a), and of the mixture of CuS and $Cu_{2-\delta}S$ (b) [437, 438]. As is seen from the figure, under certain experimental conditions CuS is capable of reduction only, and Cu_2S of oxidation only, with the reduction proceeding (in accord with [437, 438]) only to Cu_2S, and Cu_2S oxidizing, correspondingly, to CuS.

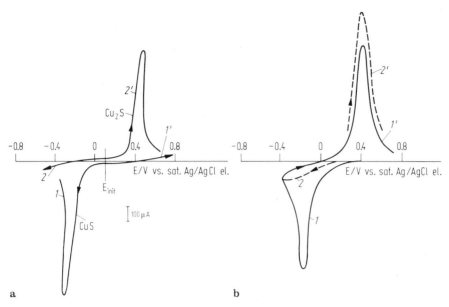

Figure 5.3. Cathodic (*1, 2*) and anodic (*1', 2'*) voltammograms of CuS (*1* and *1'*) and Cu$_2$S (*2, 2'*) registered in 0.5 M H$_2$SO$_4$ (*a*) and cathodic-anodic voltammograms of a mixture (*b*) containing 57 mass % and 43 mass % CuS (*1*) and Cu$_{2-\delta}$S (*2*) [458].

Indeed, reducing CuS during cathodic polarization (process I), the authors [458] oxidize the reduction product on subsequent anodic polarization, with the current maximum being registered at the same potential as that of the corresponding oxidation current maximum of Cu$_2$S (process II).

Here the following reactions take place [456]:

$$2CuS + 2H^+ + 2e^- \rightarrow Cu_2S + H_2S \quad \text{(process I)}$$
$$Cu_2S = Cu^{2+} + CuS + 2e^- \quad \text{(process II)}$$

The mole fraction α of Cu^{2+} in the mixture is calculated from the ratio of the maximum cathodic and anodic currents and the percentage x of CuS [437] is determined by the formula

$$x = \frac{2\alpha M_{CuS}\, 10}{2\alpha M_{CuS} + (1-\alpha)\, M_{Cu_2S}} \tag{5.1}$$

where $\alpha = q_c/q_a$ is the ratio of the electricity quantities consumed during cathodic and anodic processes, and M is the molar mass.

The value of δ in nonstoichiometric sulfide $Cu_{2-\delta}S$ can be calculated similarly [437], allowing for the relation $Cu_2S = Cu_{2-\delta}S + \delta/2$ CuS.

Applicability of voltammetry to the phase analysis of silver sulfide [459] containing metallic silver as an impurity is due to the possibility of separating electrochemical processes associated with transformations of the metallic silver and silver sulfide. Over the potential interval of about $+0.4$ V only metallic silver exhibits electrochemical activity, whereas silver sulfide remains electrochemically passive. The value of the silver oxidation current maximum is proportional to the content of silver metal in the paste. This allows us to use the given signal as the one indicating the presence of metallic silver in a sample. The voltammogram exhibits two maxima corresponding to oxidation of silver and silver sulfide.

Maxima of the reduction of $CdCN_2$ and CdS [460] are registered at rather different potentials. The value of the cadmium cyanamide reduction maximum is directly proportional to the compound content in the paste. To eliminate the effect of dispersity and possible nonuniform distribution of the substance in the paste, it is expedient to use the ratio of the heights of $CdCN_2$ and CdS peaks as the response. In this case, the determination error does not exceed 3.8% for small amounts and 2.0% for large amounts of cadmium cyanamide.

Consequently, not the absolute values of the currents but their ratio is used as indicator signals, a fact which makes it possible to introduce an arbitrary amount of the substance into the electrode reaction. This permits one to study not only powders but also films. The technique proposed was tested on a series of films and powders containing, as determined by the X-ray phase analysis, a cadmium cyanamide phase. Results of chemical and electrochemical analyses show satisfactory mutual agreement.

Thus, the use of CPEE SV [437, 458–463] in the study of sulfides solved the following problems: determination of the phase composition of sulfide materials (potential of the maximum in the voltammogram of the substrate depends on the substance nature, while the value of the maximum is determined by the substance amount); determination of the composition of binary solid solutions; determination of the nature and in some cases the amount of the dopant [437, 458–460]; detection of zinc sulfide compounds by the oxidation current of zinc which is deposited on the electrode surface after anodic and subsequent cathodic polarization of the electrode [463]; identification of crystalline hydrates; estimation of the reaction capacity of compounds in reactions of solid-phase synthesis.

A CPEF containing elemental tellurium exhibits cathodic ($E_M = -0.88$ V) and anodic ($E_M = -0.75$ V) peaks in 1 M HCl. The former peak corresponds to reduction of tellurium to H_2Te and the latter to oxidation to tellurite-ions [464]. The anodic peak of lead telluride is registered at a potential of $+0.58$ V;

two waves ($E_{1/2} = 0.30$ and 0.58 V) appear on the cathodic branch. As distinct from Te and TeO_2, selenium and SeO_2 do not display electrochemical activity [465]. Selenium(IV) is reduced from Na_2SeO_3 during cathodic polarization [465]. $PbSeO_3$ gives two cathodic peaks, the first corresponding to reduction of lead complicated by reduction of Se^{4+}.

Varying electrochemical behavior of the given materials allows phase analysis of tellurium and selenium-containing solid substances.

5.3. OXIDES

Phase analysis of oxides is based on reactions of the oxide reduction to metal and on electrochemical transformations of one oxide to another. These processes were used for phase analysis of rhenium compounds [466], vanadium oxides [467], copper oxides [468, 469], silver [457], and Cr^{6+} compounds [470] and for determination of metals mixed with oxides. A direct proportionality is observed as a rule between the maximum current and the content of electroactive phase in the CPEE.

The CPEE SV is used [471–474] for the investigation of the phase and elementary composition of thin oxide layers of materials used in microelectronics. In particular, electrochemical behavior exhibited at a carbon paste electrode by the compounds of the In–Sb–O system (Sb^0, α- and β-Sb_2O_3, α- and β-Sb_2O_4, Sb_6O_{13}, Sb_2O_5, In_2O_3, $InSbO_4$, InSb) [438, 471, 474] was studied. Anodic–cathodic voltammograms recorded using a CPEE containing a mixture of the corresponding substances [474] shows that a relationship between the electrochemical behavior of the compounds and their crystalline structure exists [472, 475]. Analogous data were obtained for arsenic and its oxides. Differences show up in oxidation (As) and reduction (As_2O_3) potentials, in their amorphous and crystalline forms, with various crystalline modifications clearly seen.

A thermodynamic analysis of the In–Sb–O system was performed. The CPEE SV was used to examine the phase composition of films deposited on indium antimonide. Phase diagrams were constructed. It was shown that a thin surface layer ($d = 2-4$ nm) consists of Sb_2O_3, In_2O_3, In. The layer composition depends to a great extent on annealing conditions and the nature of the gaseous medium and etchant. Thus, information was obtained which is of practical importance in the production of MDS (metal-dielectric-semiconductors) structures.

Thus, the use of a simple and very sensitive technique yields information about chemical composition of thin oxide layers, which was previously obtained usually using such complicated and expensive methods as photoelectron or Auger spectroscopy. The authors [472] used the developed

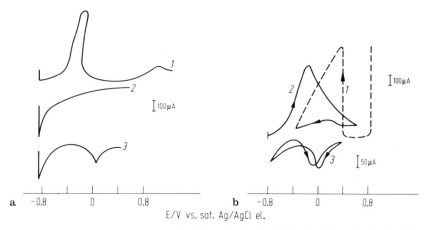

E/V vs. sat. Ag/AgCl el.

Figure 5.4. Anodic voltammograms of (a) Fe_{met} (1), Fe_2O_3 (2), Fe_3O_4 (3), and cathodic (1) and cyclic i-E voltammograms (2, 3) of (b) Fe_{met} (1, 2) and Fe_3O_4 (3) [446].

techniques also to determine the phase and elementary compositions of dielectric layers.

A CPEE was utilized [475] to investigate electrochemical behavior of hexagonal- and cubic-structure In_2O_3, $InO(OH)$, and $In(OH)_3$ of different dispersity in HCl and H_2SO_4 solutions. The reaction capacity of In^{3+} ions depends on the structure and energy state of ions in the crystal lattice.

Electrochemical behavior of tungsten compounds WO_3, H_2WO_4, Na_2WO_4, and $(NH_4)_2WO_4$ was studied. These are reduced, forming two waves or a wave and a peak in 0.5 M HCl (WO_3) and 2 M HCl (all the other compounds) [476]. $CaWO_4$ is reduced in a medium of higher acidity. Differences in the behavior of individual oxide phases MoO_3, WO_3, Eu_2O_3, $CaMoO_4$, $CaWO_4$, and $CaMo_xW_{1-x}O_4$ in 1 and 5 M HCl are exploited [477] for rapid phase analysis of, for example, molybdenum-scheelite concentrates for the content of MoO_3, $CaMoO_4$, and $CaMo_xW_{1-x}O_4$. The authors used the method of phase analysis to monitor the development of solid-phase processes in oxide mixtures $CaO-MoO_3$, $Eu_2O_3-MoO_3$, and $Eu_2O_3-WO_3$.

Figure 5.4 [446] shows voltammograms of the oxidation of iron and reduction of its oxides. Anodic oxidation of iron in a 0.1 M HCl + 0.5 M KCl solution is depicted by a voltammogram (1a) with almost symmetrical peaks corresponding to stepwise oxidation of iron to Fe^{2+} ($E_M \approx -0.3$ V) and Fe^{3+} ($E_M \approx 0.9$ V). The cathodic voltammogram of iron in acid solutions (1b) exhibits an asymmetrical anodic peak ($E_M \approx 0.1$ V) when polarization is started from a potential more positive than 0.3 V (reverse current). The cyclic voltammogram (2b) also exhibits anodic current in the cathodic potential

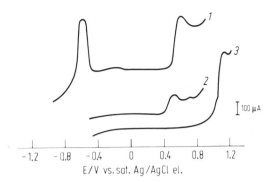

Figure 5.5. Anodic voltammograms registered using CPEE containing yellow (*1*, *2*) and red (*3*) PbO (0.1 g per 1 g of dry mixture); supporting electrolyte 0.1 M NaOH; initial potential -1.1 V (*1*), -0.6 V (*2*, *3*).

scan (reverse current). The usual and reverse anodic signals of iron fall within practically the same region of potentials. The reverse currents are probably depassivation currents.

Unlike reverse anodic currents for iron, polarization of CPEE containing Fe_3O_4 (*3b*) gives rise to reverse cathodic current in the anodic scan of the voltage. Presence of different signals for iron and its oxides allows rapid qualitative analysis of iron-ore materials and products of their processing. Thus, anodic peaks (usual $E_M \approx -0.3$ V and reverse $E_M \approx 0.1$ V) testify to the presence of metallic iron. Cathodic currents close to zero are characteristic of Fe_3O_4.

A CPEE was utilized also for phase analysis of the $Fe + FeO + Fe_3O_4 + \alpha$-$Fe_2O_3$ system [478]. Fe and FeO can be determined in the presence of any of the given components, while Fe_3O_4 can be determined only in binary mixtures with α-Fe_2O_3 [479].

Figure 5.5 shows anodic curves of lead oxides. If registration is started from a potential sufficient for reduction of oxides (*1*), the anodic voltammogram exhibits two peaks: oxidation of lead to Pb^{2+} at -0.6 V and further oxidation to Pb^{4+} at 0.6 V.

It is well known that the yellow modification of PbO and α-PbO_2 has one and the same type of crystal lattice—orthorhombic, while the red modification of PbO and β-PbO_2 exhibits the tetragonal type of crystal lattice. It is natural to expect that the process of solid substance oxidation proceeds with a minimum reconstruction of the crystal lattice. Then the processs in the region of potentials near $+0.6$ V corresponds to the transition of the orthorhombic yellow modification of PbO to the orthorhombic α-PbO_2, while the process in the region of potentials close to $+1.2$ V corresponds to oxidation of the tetragonal red modification PbO to tetragonal β-PbO$_2$. A large difference in the oxidation potentials of oxides of different modification makes it possible to determine the presence of a particular phase in the mixture.

Using a CPEE, the mechanism of cathodic and anodic electrochemical processes on several copper oxide compounds in 0.1–1.0 mol/L hydrochloric acid media has been investigated. It is shown that the cathodic reaction of CuO is much more sensitive under selected conditions than that of other copper-containing phases, which results in the variation of the preparation procedure of YBaCuO superconducting materials. A quantitative analysis is thus possible to determine the free CuO with a content of 0.5–10 wt% in the mixture with YBa$_2$Cu$_3$O$_{7-x}$ with about 10–20% relative standard deviation. The cathodic reduction reaction by chemical-electrochemical mechanisms or by solid-state conversion is also discussed [469].

A general scheme of the cathodic reduction of V$_2$O$_5$ oxide at pH = 2–5 can be represented [480] as

$$V_2O_5 + ae^- + aH^+ + pH_2O = Product + qH_2O \qquad (5.2)$$

with p or q being equal to zero. It is suggested that the products of the solid-phase reduction are V$_2$O$_4$ and oxides of the type V$_n$O$_{2n-1}$ [481, 482], vanadium hydroxide VO(OH)$_2$ [480], and solid solution of hydrogen in V$_2$O$_5$ or V$_2$O$_{5-\delta}$ ($\delta \leqslant 0.67$) [483]. According to [484], a noticeable dissolution of the cathodic reaction product is observed at pH < 2 and complications arise, which are connected with reduction of both the solid oxide and VO$_2^+$ ions, accompanied by formation of VO^{2+} and V^{3+}. Processes of the type

$$VO_2^+ + V^{3+} = 2VO^{2+} \qquad (5.3)$$

cannot be excluded either.

Different authors obtained ambiguous results from voltammetric studies of V$_2$O$_5$ dispersed in the bulk of a CPEE in solutions of phosphoric [485, 486] and sulfuric [436, 485] acids. Thus, in [436] two cathodic signals were registered: one signal at potentials 1.0 V (n.h.e.) and a weakly pronounced signal at 0.45 V. Cathodic current maxima are observed at potentials 0.45 V [485] and -0.48 V [486].

In [487] the electrochemical behavior of vanadium(V) oxide dispersed in a CPEE is studied as a function of acidity of sulfate solutions. Cathodic polarization curves of a CPEE (Fig. 5.6, solid lines) containing vanadium(V) oxide exhibit one, two, or three responses over the potential interval 0.8–0.0 V (sat. Ag/AgCl). When low-acidity solutions are used, signals 1k and 3k

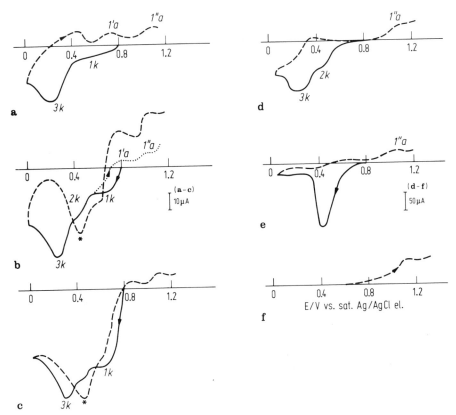

Figure 5.6. Cathodic–anodic (a–e) and anodic (f) voltammograms registered during polarization of V_2O_5-containing CPEE (a–c) and CPE (d–f): (a) 0.5 M $K_2SO_4 + H_2SO_4$, pH 2; (b) 0.5 M H_2SO_4; (c) 0.5 M $H_2SO_4 + 0.02$ M V^V; (d) 0.5 M $K_2SO_4 + M$ $H_2SO_4 + 0.1$ M V^V; pH 2; (e) 0.5 M $H_2SO_4 + 0.1$ M V^V; (f) 0.5 M $H_2SO_4 + 0.05$ M V^{IV} [487].

(Fig. 5.6) are registered. As acidity is raised, cathodic currents rise and are resolved better; intermediate signal 2k shows up, which becomes the only signal in 6 M H_2SO_4.

Current 1k rises when vanadium(V) is added to the solution (Fig. 5.6, curve c). In addition to the CPEE, current 2k can be observed in a freshly prepared vanadium(V)-containing solution at a carbon paste electrode without electroactive substance (CPE) (Fig. 5.6d, e). In this case the current is at its maximum when pH $\approx$ 1.5, lowers drastically, and vanishes with time in solutions stronger than 0.2 M with reference to H_2SO_4.

The magnitude of the signals at CPEE depends on the method used to prepare the oxide. This statement is illustrated by changes in the slope of the "current $3k$ maximum versus V_2O_5 concentration in the paste" curve. A linear dependence is observed between the current $3k$ maximum and the square root of the electrode potential scan rate.

Reduction products that are formed at potentials of current $1k$ are oxidized (Fig. 5.6, dashed lines) at potentials $1'a$ and $1''a$ (Fig. 5.6a). The maximum of the anodic current $1''a$ coincides in position with an anodic wave in the region 1.1 V, which is observed during CPE polarization in a vanadyl sulphate solution (Fig. 5.6f), and disappears when the solution is stirred. At a potential of 0.85–0.90 V the anodic current $1'a$ is more pronounced with higher contents of vanadium(V) oxide in the paste. It does not depend on stirring the solution and is due to oxidation of the reduction products localized on the electrode surface.

When the polarization direction is changed from cathodic to anodic in the region of potentials more negative than the potential of the $3k$ maximum, an anomalous [80] cathodic (reverse) [487] current (marked with an asterisk) is registered on the anodic branch of the curve. The current magnitude depends on the supporting electrolyte acidity and the concentration of V_2O_5 in the paste. Thus, when pH $\geqslant 2$, this phenomenon is not observed. The reverse current grows as acidity is increased. An extreme dependence of the reverse current on the potential scan rate, electrode holding time at the potential of signal $3k$, and the potential of the polarization reversal from cathodic to anodic was detected.

The authors [487] interpreted the data obtained, taking into account the following circumstances: the tendency of V_2O_5 to hydration and the fact that the gel $V_2O_5 \cdot nH_2O$ retains the orthorhombic structure of a solid oxide [488]; some dissolubility of vanadium(V) oxide in acid media; participation of hydrogen ions in electrochemical reduction of oxides, and the known topochemical reactions in the V_2O_5–V_4O_9–V_6O_{13} series [489]. Then, considering that current $1k$ appearing in an acid solution rises with addition of vanadium(V) to the solution and is connected with anodic signal $1''a$ which disappears in the solution being stirred, they concluded that over this potential interval one of the reduction products of V_2O_5 is represented by VO^{2+} ions which are responsible for the oxidation peak $1''a$. This conclusion is also supported by the phase diagram of the vanadium–water system.

Anodic current $1'a$ is due to oxidation of the reduction products localized on the electrode surface. In accord with [489], reduction of vanadium(V) oxide results in the formation of an oxygen nonstoichiometric oxide. Vacancies in the oxide oxygen sublattice have a random distribution. As the reduction proceeds further, anion vacancies become ordered, giving rise to a superstructure to which the V_4O_9 phase corresponds. The phase grows on the surface of vanadium(V) oxide. A structural correspondence between V_2O_5 and V_4O_9 oxides leads (see Chapter 6) to analogous behavior of chemisorbed oxygen on their surfaces. At potentials corresponding to the maximum of anodic current $1'a$ oxidation of the V_4O_9 oxide takes place. From this it may be inferred that reduction in the region of potentials of current $1k$ gives rise to a product with a considerable oxygen deficiency (V_4O_9 phase in the limiting case). Current $1k$ is probably the total current, including reduction current of oxygen absorbed on the surface of vanadium(V) oxide, since this signal increases after anodic polarization of CPEE [487].

An increase in current $1k$ in the presence of vanadium(V) in the liquid phase is likely due to the formation of highly hydrated adsorption layers and to the corresponding enlargement of the electroactive surface.

Current $2k$ is associated with reduction of VO_2^+ adsorbed from the solution or formed on the CPEE surface as a result of interaction between V_2O_5 and hydrogen ions. Then the increase of the signal at the CPEE is natural in a strongly acidic solution. For the CPE, the extreme dependence of current $2k$ on the electrolyte acidity (maximum at pH ≈ 1.5) is due to two factors: an increase in vanadium(V) oxide solubility with rising acidity and, as a consequence, lowering of adsorbability of soluble products. Coagulation of vanadium(V) compounds leads to lowering of this signal at the CPE with aging of the solution.

Signal $3k$ should be attributed to reduction of V_2O_5 via V_4O_9 to V_6O_{13}, considering that this is the procedure of chemical reduction occurring at low temperatures [489].

High concentration of vacancies in the oxygen sublattice of oxides [489], which are formed on the surface as a result of reduction, gives rise to a rapid diffusion of oxygen ions from the bulk of V_2O_5 to the surface. The diffusion character of current $3k$ is confirmed by the linear dependence of the current maximum value on the square root of the potential scan rate. In consequence, the possibility appears of subsequent additional reduction of surface layers. Then one can apprehend the appearance of reverse current during anodic scan of the potential (see Fig. 5.6b, c) and the extreme dependence of the current on some factors.

Among the factors that play a not insignificant role in the formation of the reverse current are hydration effects and hydrogen ions, which should bring about a disorder in the surface layer, thus enhancing the layer reactivity. Indeed, the reverse current grows as the solution acidity is raised. It becomes imperceptible with decreasing acidity and increasing concentration of V_2O_5 in the paste, when the diffusion processes slow down.

The scheme of electrochemical reduction of vanadium(V) oxide in acid solutions (see Fig. 5.6) can be represented as follows [487]:

$$V_2O_5 \xrightarrow[1k]{H^+,e^-} V_4O_9 \xrightarrow[3k]{H^+,e^-} V_6O_{13}$$

$$V_2O_5 \cdot nH_2O \xrightarrow[3k]{H^+,e^-} VO^{2+}$$

$$VO_{2,surf.}^+ \xrightarrow[2k]{H^+,e^-} VO^{2+}$$

Figure 5.7 displays voltammograms registered using CPEE containing one of the vanadium oxides [490]. Individual features of the oxides are more pronounced in acid solutions. The higher vanadium oxide is characterized, as was noted above, by the presence of three cathodic signals. It is incapable of oxidation, as shown by the absence of anodic signals. The V_3O_7 oxide is electrochemically inactive. A specific feature of V_4O_9 is practically the same proneness to oxidation ($E \approx 0.85$ V) and to reduction ($E \approx 0.65$ V). Besides, a less pronounced cathodic signal ($E \approx 0.2$ V) and an anodic signal ($E \approx 1.1–1.2$ V) appear. V_6O_{13} and VO_2 oxides are oxidized within the same region of potentials, but only V_6O_{13} demonstrates the ability in cathodic reduction.

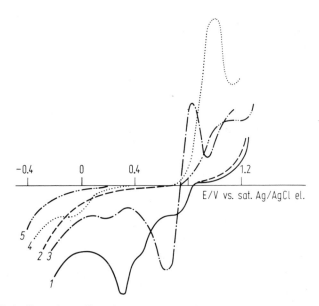

Figure 5.7. Cathodic and anodic voltammograms registered during polarization of CPEE containing: $1 — V_2O_5$; $2 — V_3O_7$; $3 — V_4O_9$; $4 — V_6O_{13}$; $5 — V_2O_4$ ($v = 20$ mV/s; 0.5 M H_2SO_4) [490].

Table 5.1. Current Maximum Potentials (s.c.e.) of Vanadium Oxide Electrochemical Transformations

Oxide	Supporting Electrolyte	Anodic Signals at v(mV/s)		Cathodic Signals at v(mV/s)			Ref.
		20	3.3	80	20	3.3	
V_2O_5	1 M HCl					0.20	485
	1 M H_2SO_4					0.25	485
	1 M H_3PO_4					−0.65	485
	0.5 M H_2SO_4			0.65	0.65		486
				0.30*	0.40		487
						0.30*	487
V_4O_9	0.5 M H_2SO_4	0.85*			0.65*		487
		1.10			0.20		490
V_6O_{13}	0.5 M H_2SO_4	1.00*		0.40	0.00*		490
				0.20			490
				0.00*			490
V_2O_4	0.5 M H_2SO_4	1.10*					490
	1 M KNO_3		0.65				490
			0.90				485
	0.05 M H_2SO_4		0.90				486
	1 M H_3PO_4				−0.87		485
V_2O_3	1 M HCl		0.60				486
	1 M H_2SO_4		0.60				485
	1 M H_3PO_4		0.55				485
	1 M KNO_3		0.70				485
	1 M NaOH		0.10				485

As the potential scan rate is increased, the oxide oxidation currents rise and are shifted to the region of positive potentials. A linear dependence is observed in the E_M–$\lg v$ coordinates with the slope equal to 0.14. The linear relation between the oxidation currents and the square root of the potential scan rate is indicative of the diffusion nature of the processes. The maximum current value changes little if at all when a CPEE containing VO_2 and V_6O_{13} is held for 5 min in 0.5 M H_2SO_4 before recording a curve, that is, the rate of chemical dissolution of the oxides from the paste is less than the electrochemical transformation rate. A direct relationship exists between the anodic current and the content of electrochemically acitve oxides in the paste.

In Table 5.1, an asterisk is used to mark the signals which can be utilized as indicator signals when developing techniques of phase analysis of the oxides comprising the V_2O_5–V_2O_4 series. The phase analysis of minerals for the content of V^{5+} and V^{4+} involves recording of the oxidation peak of V^{4+} in 1 M KNO_3 and reduction peaks of V^{5+} and V^{4+} in 1 M H_3PO_4.

Electrochemical behavior of the V_2O_3 oxide was described by Songina

[467]. It was shown that in an acid medium the oxide is oxidized, the potential of the anodic current maximum is close to $+0.50 - 0.55$ V (s.c.e.), and the process is given by

$$V_2O_3 + 2H^+ = 2VO^{2+} + H_2O + 2e^- \qquad (5.4)$$

In an alkaline medium the anodic current maximum is observed at a potential of $+0.05$ V (s.c.e.). Oxidation proceeds according to the reaction

$$V_2O_3 + 5H_2O = 2VO_4^{3-} + 1OH^+ + 4e^- \qquad (5.5)$$

Keeping the electrode in acid and alkaline solutions leads to a lowering of the maximum current owing to partial dissolution of the oxide and depletion of the surface layer of the paste. The duration of the electrode contact with a neutral solution has no influence on the signal. The current maximum is proportional to the concentration of V_2O_3 in the paste. In an acid medium, over the potential interval where this oxide is oxidized, other oxides (see Table 5.1) do not give signals. This fact allowed the authors to propose a procedure for the determination of the V_2O_3 concentration in the presence of higher vanadium oxides.

CHAPTER

6

INVESTIGATION OF SPECIFIC FEATURES OF THE STRUCTURE OF A SOLID

The properties of a solid depend on its structure (crystalline and electronic) and defects (nonstoichiometry, presence of impurities, vacancies, and interstitial atoms) which, in turn, determines the nature and mobility of current carriers, their localization, forbidden band width, and the Fermi level position. The character of electrochemical transformations at the solid–electrolyte solution interface is governed by the relative position of the energy levels of the redox system and electrons in the conduction band and valence band of a semiconductor. In addition, a substantial contribution to the electrode process is made by the potential drop in the space charge region in a semiconductor at the phase boundary and by the concentration of free electron carriers that influence the adsorption processes, participate in electrode reactions, and determine the reaction rate. The ratio of the energies of the electrons in a semiconductor (Fermi level) and those in the redox system in the solution determines the thermo-dynamics of the electrode process, while the potential drop in the space charge region and the concentration of current carriers define the electrode process kinetics.

With this in mind it is reasonable to pose a converse problem; searching for electrochemical responses to a specific type of disorder in a solid, establishing correlations between the electrochemical behavior of a solid and its structure, both crystalline and electronics.

Information is furnished by the equilibrium (or stationary) potential and potentiodynamic curves recorded during electrochemical transformations of the compounds studied or adsorptive layers (mainly oxygen) on the compound surface. The versatile techniques of the experiment allow the study of solid substances in the form of single crystals, polycrystalline specimens, pyrolytic films, and powders dispersed in a carbon paste electrode or applied to the electrode surface.

6.1. EQUILIBRIUM AND STATIONARY POTENTIALS

As stated in [491], under equilibrium conditions a potential that depends on the degree of nonstoichiometry of the substance constituting the covering layer is set up between the layer anions and the electrolyte at the interface of the electrolyte and the nonstoichiometric covering layer on the current conductor featuring electronic conductivity.

For an oxide covering layer which is characterized by the equilibrium O^{2-} (covering layer) $+ 2H^+_{ads} = H_2O$, one can write

$$E = \frac{1}{2F}\frac{d\Delta G(n)}{dn} + E^0_{O_2} + \frac{RT}{F}\ln a_{H^+} \tag{6.1}$$

where n is the number of oxygen atoms per one metal atom in the oxide; the quantity $[1/(2F)]\,[d\Delta G(n)/(dn)]$ is the electrode potential of the oxide electrode as referred to the oxygen electrode in the same solution; $E^0_{O_2} = 1.227\ \text{V}$ is the standard potential of the oxygen electrode.

Under conditions of cation and anion equilibrium, a potential sets up, satisfying a certain composition of the nonstoichiometric layer ($n = n^*$).

Vetter and Jeger [492] investigated the electrode potential of γ-manganese dioxide deposited at the platinum anode as a function of the Mn^{2+} ion concentration, pH, and n. Over the region of homogeneity of the MnO_n compound ($1.5 \leqslant n \leqslant 2$), they observed a variation of the potential with time. The initial potential values were determined by the starting value of n, and subsequently a potential corresponding to MnO_{n*} (a compound which is in equilibrium with the electrolyte of preset composition) sets up.

Using Honig's model, according to which the silver halide lattice can be considered as a solvent and vacancies and interstitial ions can be assumed to be dissolved anions and cations, Hoffman obtained an expression for the standard potential in the form

$$E^0 = \frac{RT}{nF}\ln\frac{1}{N_i} + D \tag{6.2}$$

where N_i is the concentration of the interstitial Ag_i^+ ions and D is a constant.

Taking silver halides with a given defect as an example, he showed that Eq. 6.2 agreed with the experiment. The author [493] also observed a shift of the standard potential of silver bromide containing $CdBr_2$ toward positive values, corresponding to a decrease in the concentration of interstitial silver ions.

Usually it is difficult to distinguish between equilibrium and stationary processes at the semiconductor (or insulator)–electrolyte solution interface. However, as was shown experimentally, the potential set up depends on the nature and nonstoichiometry of the compound [494]. Hence most authors use the notion of the stationary potential (E_{st}). We will also use this term further in the text.

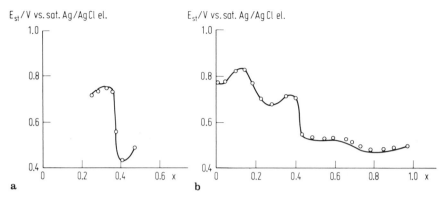

Figure 6.1. Stationary potential versus composition of sodium (*a*) [495] and silver (*b*) [496] vanadium oxide bronzes.

A regular change of the stationary potential was revealed over the region of homogeneity of sodium [495] and silver [496] vanadium bronzes (Fig. 6.1), in the V_2O_5–MoO_3 system, over wide limits of variation δ in the $Zn_{1-x}Sr_x$ $CoO_{3-\delta}$ compounds [497]. In $Ca_3(V_{1-x}P_xO_4)_2$ and $Sr_{3(1-x)}Ba_{3x}(VO_4)_2$ solid solutions isovalent substitution is accompanied by a drop of the stationary potential of a carbon paste electrode containing the corresponding solid compound with increasing x:

x in $Ca_3(V_{1-x}P_xO_4)_2$	0.01	0.1	0.2	0.4	0.5	1.0
$E_{st} \pm 0.02$ V (sat.Ag/AgCl)	0.85	0.86	0.85	0.85	0.88	0.49

A sharp drop in the stationary potential in the series of solid solutions of the $CuFe_2O_4$–$Cu_{0.5}Fe_{2.5}O_4$–Fe_3O_4 system was observed in [498]. The authors believed that the change in the redox properties is due to a change in the nature of the solid solution.

Lead sulfide with controlled deviation from the composition stoichiometry was taken as an example [461] to show variations of the stationary potential of a carbon paste electrode containing PbS dispersed in the bulk of the electrode within the limits of the compounds homogeneity. A maximum E_{st} corresponds to stoichiometric PbS; E_{st} lowers as the sample composition deviates from stoichiometry.

Thus, the equilibrium (or stationary) potential that sets up at the solid–electrolyte solution interface is sensitive to the deviation of the composition from stoichiometry and changes regularly as the composition varies within the limits of the compound homogeneity.

6.2. TYPICAL ELECTROCHEMICAL TRANSFORMATIONS OF SOLIDS

A relationship between the character of electrochemical transformations and the crystalline structure type and defects was detected for some systems. In their study of electroreduction of halogenides dispersed in a CPEE, Ruby and Tremmel [499] suggested that interstitial silver ions enter into the reaction; the loss of the ions is replenished by a rapid transition of ions from regular sites to interstitial sites. Surplus anions go into solution.

It is common knowledge that there are different potentials of redox transformations in the systems of α- and β-modifications of lead oxides, α-, β-, and γ-modifications of manganese oxides [500], oxidation of crystalline and amorphous arsenic and its oxides, and antimony oxides of various modifications which were discussed in Section 5.3. It is possible to simply and rapidly evaluate the degree of ordering in the crystalline structure of CdS from the shift of the compound reduction potential toward negative values [460].

A higher-temperature synthesis or anodic polarization enhances the stability of oxides, a fact which is related to the structure ordering and partially to a decrease in the specific surface. A lower stability of spinel oxides compared to the initial simple oxides is thought to associated with a higher level of structure defects of the former. Redox properties of nonstoichiometric oxides of the series V_2O_5, V_3O_7, V_4O_9, V_6O_{13}, VO_2 were studied in [487, 490]. The redox properties of the oxides do not change smoothly in the given series. Subject to cathodic polarization, transformations in the series V_2O_5, V_4O_9, V_6O_{13} proceed rather easily, a fact which agrees with the results obtained for thermal reduction of oxides [489] and which is due to the closeness of the structures. The V_3O_7 oxide exhibiting a different structure falls out of the V_2O_5–VO_2 oxide series considered. Electrochemical activity is not characteristic of V_3O_7, although it is next to the higher oxide V_2O_5 in the given series. Resistance of the V_3O_7 oxide to redox processes is explained probably by its structure, which is different from the structure of the oxides given above, by a more ordered distribution of d-electrons, and by lower electroconductivity compared to V_4O_9 and V_6O_{13}. A decrease in the oxidation current in going from V_6O_{13} to V_2O_4 also correlates with lower electroconductivity of the oxides.

An interesting change of electrochemical activity in the series of sodium vanadium oxide bronzes over the homogeneity interval of the δ-phase was observed by Bazarova [495]. Oxidizability of the bronzes depends on the composition (Fig. 6.2). The $Na_{0.33}V_2O_5$ bronze features a minimum electrochemical activity. Annealing of this bronze in vacuum is accompanied by the appearance of a well-defined anodic signal on the voltammogram at

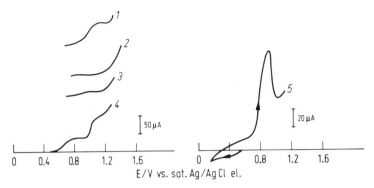

E/V vs. sat. Ag/AgCl el.

Figure 6.2. Anodic (*1–4*) and cathodic–anodic (*5*) voltammograms of sodium vanadium oxide bronzes $Na_xV_2O_5$. $x = 0.22$ (*1*); 0.33 (*2, 5*); 0.35 (*3*); 0.40 (*4*) before (*1–4*) and after (*5*) annealing in vacuum. CPEE composition: 0.1 g (*1–4*) and 0.2 g (*5*). $Na_xV_2O_5$ in 1 g of dry mixture [495].

potentials 0.8–0.9 V. A similar effect was observed [490] for annealing V_2O_5 in a vacuum. This anodic current may be due either to the electronic disorder resulting from oxygen loss or to the emergence of the phase of a lower oxide capable of oxidation. Unfortunately, available data are insufficient for solving this problem. It is known however that electroconductivity of samples rises sharply after annealing of the bronze in vacuum. This suggests realization of the former mechanism of the anodic current generation, because X-ray phase analysis does not reveal the appearance of a new phase.

It follows from the given data that a minimum electrochemical activity is shown by simple and complex oxides exhibiting a maximum ordered structure. Disordering of the compound lattice, which is accompanied by electronic disordering (electroconductivity growth), results in higher electrochemical activity of the compound.

A rather interesting study of simple and complex iron oxides was made by Lecuire and Evrard [501]. They found a correlation between the character of voltammograms and position of the Fe^{3+} ions at tetra- and octahedral sites. Figure 6.3 displays the voltammograms. The authors attribute signal I to reduction of Fe^{3+} from tetrahedral sites and signal II to reduction of Fe^{3+} ions from octahedral sites.

A rotating gold-ring CPEE containing various iron oxides (α-Fe_2O_3, γ-Fe_2O_3, Fe_3O_4, $Fe_{1-\delta}O$) was used [5O2] to determine the nonstoichiometry of oxides. Cathodic reduction of the oxide from the CPEE is accompanied by appearance of Fe^{2+} oxidation current at the ring. Assuming that the amount of electricity during the reduction process depends only on Fe^{3+} reduction, the authors calculated the content of Fe^{3+} in the oxides studied and the ratio $Fe^{3+} : Fe^{2+}$ using coulometric data. The authors proposed that

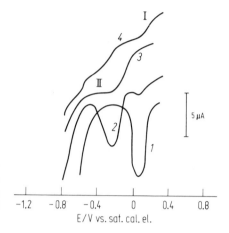

Figure 6.3. Cathodic voltammograms registered during polarization of carbon-paste electrode with a conductive binder containing iron oxides Fe_3O_4 (*1*), FeO (*2*), α-Fe_2O_3 (*3*), and γ-Fe_2O_3 (*4*) (1 M $HClO_4$; $v = 60$ mV/min) [501].

the method be used for checking the variation of composition with depth in layers of nonhomogeneous surface oxides.

Lecuire determined deviation from stoichiometry in $Fe_{1-\delta}O$ from the value of Fe^{3+} reduction current registered at the electrode containing an electroconductive binder.

Pure and doped lead sulfide was taken as an example [461] to show the possibility of using typical electrochemical transformations of the substance to be investigated in the study of nonstoichiometry of binary chalcogenides. The lattice defectness of the test samples was preset by changing thermodynamic parameters of the medium (temperature, partial pressure of sulfur vapors) and by doping with heterovalent impurities of acceptor (Na, Ag) and donor (Cl) types.

Figure 6.4 presents anodic voltammograms recorded after cathodic polarization of a CPE in a solution of 1.7 M NaOH and 1×10^{-3} M $Pb(CH_3COO)_2$, and of a PbS-containing CPEE in a 1.7-M NaOH solution. A comparison of the curves shows that the electrochemical signals on curve *2* correspond to oxidation processes of lead and, probably, sulfide sulfur.

It follows from the temperature dependence of the ratio of the maximum oxidation currents of lead and sulfur in samples annealed in an argon stream that i_{Pb}/i_S lowers as the temperature is raised. This is indicative of the sulfur content increasing relative to the lead content. A minimum ratio of the oxidation currents is registered in the region of stoichiometric PbS, where the electroneutrality condition becomes $n = p$. The authors found a correlation between the ratio of lead and sulfur oxidation currents, concentration of current carriers in a semiconductor, and the nature of disorder preset by changing thermodynamic parameters of the medium or by introducing isovalent impurities.

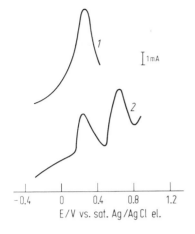

-0.4   0   0.4   0.8   1.2

E/V vs. sat. Ag/AgCl el.

Figure 6.4. Anodic voltammograms registered using CPE (*1*) and PbS-containing CPEE (*2*) (The curves were recorded from a potential -0.8 V to $+1.0$ V [461]).

Sapozhnikova and Roizenblat [503] developed the concepts of the vacancy mechanism of solid-phase electrochemical reactions. Examining electrochemical transformations of the lower copper oxide Cu_2O and iron [503] and nickel [504] oxides, they showed that those substances were capable of electrochemical reduction, with anodic reactions being pronounced much weaker. The authors think that this is due to participation of vacancies in electrode processes and formation of intermediate neutral associates which, on passing into solution, give Cu^+ and Fe^{2+} ions. Proceeding from these considerations, the authors interpreted results of investigations concerned with reduction of α- and γ-Fe_2O_3, magnitite, wustite, and barium ferrite. The authors [503] ascribe the different behavior of α- and γ-Fe_2O_3 (electrochemical inactivity of the former oxide and activity of the latter) to the fact that γ-Fe_sO_3 has cation vacancies which form associates with the metal ion, while α-Fe_2O_3 has no such vacancies. As reported by Lecuire [502], both modifications of Fe_2O_3 are capable of reduction: α-Fe_2O_3 is reduced at a potential ~ 0.1 V and γ-Fe_2O_3 is reduced in two stages ($E_{1/2} \sim 0.4$ and -0.1 V). A higher electrochemical activity of γ-Fe_2O_3 is explained by a lower binding energy of Fe^{3+} at tetrahedral sites than at octahedral sites.

The degree of cation oxidation in $(Fe_3O_4)_x(CuFe_2O_4)_{1-x}$ and $[Li_{0.5}Fe_{2.5}O_4]_y[(Fe_3O_4)_x(CuFe_2O_4)_{1-x}]_{1-y}$ solid solutions was studied [498, 500]. Oxidation currents of Fe^{2+} and Cu^+ ions and reduction currents of Fe^{3+} were registered depending on the value of x. Sharp variations of the currents were observed at $x = 0.5$, a fact which the authors attribute to a change in the nature of the solid solution, assuming that at $0 \leqslant x \leqslant 0.5$ the solid solution composition can be represented as $Cu_x^+ Cu_{1-2x}^{2+} Fe_{2-x}^{3+} O_4$, while at $x > 0.5$ as $Cu_{1-x}^+ Fe_{2x-1}^{2+} Fe_{3-x}^{3+} O_4$. The content of Cu^+ and Fe^{3+} ions is at a maximum when $x = 0.5$.

Cu^+ ions entering into the ferrite composition and forming the Cu_2O phase are distinguished electrochemically [500]: a maximum oxidation current of the Cu_2O phase is observed at a potential close to $+0.3$ V (sat.Ag/AgCl). When studying ferrites, the authors registered an anodic current over this potential interval only after oxidation of the Fe^{2+} ions.

Thus, typical electrochemical transformations of solids give useful information about the phase composition, type, and ordering of the compound crystal lattice and in some cases can be used to determine nonstoichiometry and disorder of the solids structure. However, a paucity of experimental data and ambiguous interpretation of results calls for special caution in drawing conclusions about the compound structure defects and require in some cases the use of independent methods of investigation.

6.3. CHEMISORPTION OF OXYGEN ON OXIDE SURFACES— FORMATION OF SIGNALS OF THE FIRST AND SECOND KINDS

Much attention has been paid recently [21, 435, 505] to the possibility of using electrochemical transformations of chemisorbed oxygen as a source of information about the structural defects of solids. Note that in the literature the term "chemisorbed oxygen" is encountered as often as "adsorbed oxygen." Being unable to draw a clear distinction between the terms, we will use both terms, as was done by the authors of the papers cited.

An oxygen molecule or atom possesses sufficient affinity for electrons so as to capture or to localize them to form O_2^- or O^- particles. The charge that compensates for the charge of an ionized adsorbed atom or molecule can be localized at the lattice defect well apart from the adsorbate particle or can be "smeared" over the crystal surface (free electrons in the conduction band). This mechanism of oxygen adsorption is typical of n-type semiconductors. In this case oxygen adsorption is followed by formation of a depleted space-charge layer. Oxygen adsorption on p-type semiconductors leads to formation of an enriched space-charge layer. Indeed, oxygen adsorption on NiO, Cu_2O is accompanied by a rise of the surface conductivity owing to formation of $O_2^{\delta-}$ or $O^{\delta-}$ particles and appearance of a hole in the surface layer.

For ionic semiconductors and dielectrics, adsorption proceeds mainly with the participation of locally acting forces. The adsorbate becomes linked with a surface atom or group of atoms. Here the adsorbate interacts with surface states of "broken bonds" (unpaired electrons). The local interaction between an adsorbed atom and the surface may be also due to dipole–dipole attraction. Covalent bonds of O^- and O_2^- with the surface are formed if the latter has donor centers (metal cations, Lewis acids, anion vacancies) [506, p. 115]. The donor–acceptor mechanism of adsorption represents probably the most general case [506, p. 127]. These concepts rest on the assumption that adsorption takes place at surface defects. Electron-acceptor (EA) centers are represented by surface coordinate-unsaturated atoms of metal M, whose effective charges q_{eff} differ from the charges of regular atoms in the bulk of the crystal. Vacancy-type defects often serve as electron-donor (ED) centers. Adsorption is accompanied by formation of complexes of the $A^{\delta-} M^{q_{eff}\delta+}$ or $D^{\delta+} M^{q_{eff}\delta-}$ type (A and D being the acceptor and donor particles of the adsorbate, δ the excess charge) [506, p. 128]. In this case carriers are not necessarily localized on the complex orbitals and the maximum of the wave function of the complex can be localized, for example, at a neighboring defect. Any type of interaction is possible here: from ion-dipole to partially covalent and ionic interactions, as well as the formation of a charge-transfer complex (CTC).

Thus, oxygen chemisorption centers may be vacancies in cation and anion sublattices, quasi-free electrons, and ions of transition metals with the lowest degree of oxidation [497]. Then it is possible to determine the defect structure of an oxide using data on oxygen adsorption on the oxide surface. Some studies have been made recently in this direction. Two approahes aimed at solving this problem have been taken, which differ in the choice of the indicator signal. In one case, use is made of the reduction current of oxygen chemisorbed from the gaseous phase or solution [505, 507]. This signal is called a signal of the first kind [21, 505]. It is observed as a rule over the potential interval close to 0.0 V (sat.Ag/AgCl) [505]. In the other case, electrochemical transformations of chemisorbed products of the anodic reaction are considered and corresponding cathodic and anodic currents are registered [495, 496, 508–511]. These signals are referred to as signals of the second kind [21, 505]. These are observed over the potential interval 0.8–0.6 V (sat.Ag/AgCl). Since the cathodic process involves the anodic reaction products, the maximum cathodic current is proportional to be anodic current value, electrolysis time, and adsorbability of the solid studied.

Thus, formation of a correct signal of the first kind requires control over the partial pressure of oxygen, while electrolysis conditions are essential for signals of the second kind. Advantages of the signals of the second kind consist of a higher filling of the surface with adsorbate and the much greater values of these signals compared to signals of the first kind, as well as the possibility of adjusting exactly the effective partial pressure of oxygen, presetting a certain potential or current.

Processes of Anodic Oxidation of Water. These were considered in a series of reviews, out of which the most up-to-date and complete are, in our opinion, the works by Trasatti and Lodi [497] and Tarasevich and Efremov [512].

Evolution of oxygen on oxides is represented, for example, as follows:

1. A mechanism involving formation of a higher oxide [512, p. 245]:

$$MO + OH^- \rightarrow MOOH_{ads} + e'$$

$$MOOH_{ads} + OH^- \rightarrow MOO + H_2O + e'$$

$$MOO + OH^- \rightarrow MOOOH_{ads} + e' \qquad (6.3)$$

$$MOOOH_{ads} + OH^- \rightarrow MO + O_2 + H_2O + e'$$

2. Evolution of oxygen on $La_{0.5}Ba_{0.5}CoO_3$ [513]:

$$OH^- \rightleftarrows OH_{ads} + e'$$

$$OH^- + OH_{ads} \rightleftarrows O_{ads}^- + H_2O$$

$$O_{ads}^- \rightleftarrows O_{ads} + e' \qquad (6.4)$$

$$2O_{ads} \rightarrow O_2$$

A similar mechanism of oxygen evolution on IrO_2 in an acid medium is proposed in [514].

A mechanism of oxygen evolution on a compact Co_3O_4 was discussed proceeding from semiconductor properties of the oxide [515]. The authors think that the larger portion of the potential drop corresponds to the space charge region of the semiconductor and the water oxidation reaction rate is determined by the concentration of holes on the surface. It was found [515] that the oxygen evolution rate increased symbatically with increasing non-stoichiometry of the compound. The process is presented as follows [515]:

$$(Co_3O_4)_{surf} + 4h^{\cdot} + 2H_2O \rightarrow (Co_3O_4)_{surf} + 4H^+ + O_2 \qquad (6.5)$$

where $h^{\cdot}$ is a hole.

Taking into account the acceleration of electrode processes on the Co_3O_4 surface with rising concentration of excess oxygen, the authors came to the conclusion that defects could participate directly in electron transfer. They believe [515] that the degree of nonstoichiometry that determines the number of cation vacancies affects not only the bulk conductivity but also polarization characteristics connected with electrocatalytic activity of the oxide surface. Hopping conductivity is realized in spinel oxides. This conductivity mechanism is due to the presence of similar-type cations of different degrees of oxidation at the same crystallographic sites, a situation which can be schematically represented as

$$Co^{4+} \rightleftarrows Co^{3+} + h^{\cdot} \qquad (6.6)$$

In [477], two types of specimens with n- and p-type conductivity were investigated to find a correlation between current–voltage characteristics and the structure defectness of semiconductor oxide materials.

Figure 6.5 displays some anodic voltammograms registered from the stationary potential using CPEE containing some oxide compounds. As is seen, overvoltage of the anodic process at the p-type semiconductor (NiO) is considerably less than at the n-type semiconductors. A study of solid solutions of lithium oxide in nickel oxide, where a disorder of the type

$$Li_2O + 1/2O_2 \rightarrow 2Li'_{Ni} + 2O_0^x + 2h^{\cdot} \qquad (6.7)$$

(where x, the prime, and the dot stand for the neutral, negative-charge, and positive-charge centers, respectively) is observed, showed that the anodic process overvoltage dropped dramatically as the concentration of quasi-free holes increased. For the n-type semiconductor MoO_{3-y}, an increase of y (growth of the concentration of quasi-free electrons) results in growing overvoltage of the anodic process. It follows from the studies of the redox processes taking place at the vanadium oxide [490] (orthovanadates of alkaline-earth metals)–electrolyte solution interface that the presence of localized electrons in an n-type semiconductor encourages the rise of the anodic process overvoltage to a much lesser degree than the presence of quasi-free electrons does.

Of interest is the fact that a rather large decrease in the overvoltage of

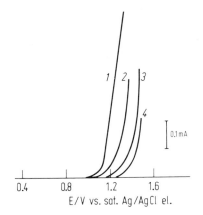

Figure 6.5. Anodic voltammograms registered during polarization of CPEE containing NiO (1), Nb_2O_5 (2), TiO_2 (3), ZnO (4) (0.5 M H_2SO_4; $E_i = +0.6$ V; $v = 0.08$ V/s).

the hydrogen evolution reaction was observed on n-type semiconductors [516, 517] when the concentration of quasi-free electrons rose.

Two important conclusions can be drawn from the foregoing: (1) water oxidation overvoltage depends on the oxide structure defects (the overvoltage drops as the concentration of quasi-free holes increases); (2) intermediate products that are formed during electrochemical evolution of oxygen are adsorbed particles of O_{ads} type or complexes of MOO type. It is suggested [518] that an extensive chemisorption of atomic oxygen is observed in an acid medium, while in alkaline solutions oxygen is released principally in the molecular form.

At a certain cathodic potential, chemisorbed oxygen is capable of localizing an electron, thus completing structural sites in the anionic sublattice. The negative surface charge arising therefrom can be compensated for by hydrogen ions that forms the outer sheath of the double electric layer. It is suggested [477] that the cathodic process involves formation of a suboxide complex on the surface of the solid phase

$$M_M^x + O_{ads} + e' \rightarrow MO'_{surf} \tag{6.8}$$

where M_M^x is the metal at its site in the crystal lattice. Oxygen of the suboxide complex can be reduced or oxidized:

$$MO'_{surf} + 2H^+ + e' \rightarrow H_2O + M_M^x \tag{6.9}$$

$$MO'_{surf} \rightarrow 1/2O_2 + e' + M_M^x \tag{6.10}$$

Indeed, cyclic curves, which were registered subsequent to sufficient anodic polarization, exhibit cathodic currents and, related to them, anodic currents.

Since the indicator signal of the second kind is caused by reduction of the anodic reaction products accumulated on the surface, it is a function of two variables: the anodic process overvoltage (current flowing at a given potential) and adsorbability of the solid, with all other things being equal. This fact makes it possible, on the one hand, to assess changes in the concentration of quasi-free electron carriers or the degree of nonstoichiometry by measuring the anodic current at a fixed potential in a series of similar-type nonstoichiometric compounds or solid solutions and, on the other hand, to impose certain requirements upon experimental procedures aimed at evaluating the adsorption ability. In the latter case, anodic polarization should be carried out probably at a preset current rather than at a present potential.

Let us consider the relationship between the signals and specific features of the compound structure.

6.3.1. The Relationship Between the Signals
and Disorder of the Crystal Lattice

The relationship between the reduction currents of chemisorbed oxygen and the compound structure defects was first clearly demonstrated in [519]. The authors examined electrochemical reduction of oxygen adsorbed on the surface of zinc oxide and ZnO doped with gallium oxide. It turned out that the quantity of electricity required for reduction of chemisorbed oxygen depends linearly on the logarithm of the quasi-free electron concentration in the samples studied. The last circumstance testifies to the existence of a correlation between the concentration of quasi-free electrons in the sample and the amount of oxygen chemisorbed on the sample surface and supports experimentally the participation of quasi-free electrons of zinc oxide in the adsorption process. This is also confirmed by an increase in the maximum reduction current of chemisorbed oxygen as the sample electroconductivity rises. The authors [520, 521] observed signals of the first kind on titanium oxides in 0.5 M H_2SO_4 over the potential interval 0.4–0.2 V (n.h.e.).

Taking into account the charge curves of anatase and rutile, the same authors came to the conclusion that the crystal structure does not play a decisive role here. The curves depend on the sample defectness (degree of nonstoichiometry, presence of reduced forms of titanium, oxygen vacancies).

Making good use of the technique proposed by Barikov and Songina [109] and Ruby and Tremmel [499], and modifying it somewhat subsequently, Sapozhnikova and Roizenblat with co-workers [504, 507, 521, 522] used signals of chemisorbed oxygen (signals of the first kind) as indicator signals relative to certain-type defects occurring in the structure of solids. Thus, changes in the signal of the first kind were observed [521] on TiO_2 and titanium oxide doped with La_2O_3 and Nb_2O_5 depending on the sample production conditions and the nature of dopant. Maximum effects were registered on samples doped with La_2O_3. The authors deduce that a relationship exists between the signal and the concentration of oxygen vacancies, assuming that the concentration is lowered when titanium oxide is doped with niobium(V) oxide and rises if TiO_2 is doped with lanthanum oxide. The explanation proposed is not conclusive, since the authors disregarded the possibility of the disordering proceeding by other mechanisms, for example, according to the reaction $La_2O_3 \rightleftarrows 2La_i^{\cdots} + 6e' + 3/2O_2$ as a result of which the concentration of quasi-free electrons and not of oxygen vacancies increases. Then the higher signal of chemisorbed oxygen, which was observed in [521], lends itself to an adequate interpretation in terms of Ref. [519].

As an additional proof of their hypothesis the authors [507] adduce the increase in the signal of the first kind when ZrO_2 is doped with calcium oxide, this resulting in disordering according to the reaction CaO →

$Ca''_{Zr} + O^x_0 + V^{..}_0$. It should be noted however that the approach adopted in [507] fails to account for the results of the paper [523] which describes electrochemical activity of the surface oxygen of nickel oxide doped with lithium oxide. Allowing for the reaction of the solid solution formation (Eq. 6.7) it is hardly possible to trace a relationship between the cathodic signal and the concentration of anion vacancies which are not present in these samples. So, the statement made in [507, 521] on the decisive role of anion vacancies of any origin in generation of the given signal calls for refinement. The corresponding data will be presented below.

At the same time, as reasonably noted by the authors [520], a certain relation exists between the degree of sample defectness (degree of nonstoichiometry, presence of reduced forms, oxygen vacancies, etc.) and oxygen adsorbability. In this respect, of interest is the variation in the character of the composition dependence of the chemisorbed oxygen reduction current (signal of the first kind) within and outside the homogeneity region of the solid solution found in the $BaO-TiO_2$ system [522] and the growth of the signal of the first kind as the TiO_2 annealing temperature [507] is raised, leading to the appearance of "donor" centers—oxygen vacancies.

The authors [510, 511] registered cyclic voltammograms from the starting potential $+0.4$ V in the anodic direction and from $+1.25$ V (sat.Ag/AgCl) in the cathodic direction. Voltammograms of p-type semiconductors NiO, $NiO \cdot xLiO_2$, ZrO_2, and Eu_2O_3 exhibited cathodic current maxima at a potential of $+0.75$ to $+0.70$ V, but they appeared only after a sufficiently long anodic polarization (signals of the second kind). The value of these signals varies symbatically with changes in the concentration of quasi-free holes, which was present in the case of NiO by doping the oxide with lithium oxide. Those current maxima were an order of magnitude lower in voltammograms of n-type semiconductor oxides (MoO_3, WO_3, ZnO, Nb_2O_5, TiO_2). The relationship between signals of the second kind and the concentration of quasi-free holes is probably due to a decrease in the anodic process overvoltage. Since all samples were polarized over the same potential interval, the sample, where the anodic reaction overvoltage was lower, accumulated more oxidation products. Thus, an indirect dependence of the signal of the second kind of the semiconductor conductivity type and the concentration of electronic current carriers was observed.

Figure 6.6 shows anodic and cathodic curves recorded after anodic polarization of vanadium(V) oxide samples annealed in a vacuum and in air. Annealing in a vacuum is accompanied by the appearance of anodic current, a fact which indicates the emergence of V^{4+} centers during annealing. In this case the cathodic signal is considerably larger than that observed on the sample annealed in air and containing no significant amounts of V^{4+} (the anodic branch of the curve exhibits no corresponding effects). The relation

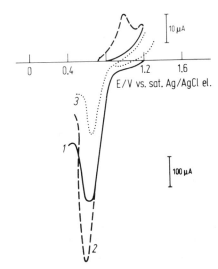

Figure 6.6. Cathodic (*1–3*) and anodic voltammograms registered during polarization of CPEE containing V_2O_5 not annealed (*1*) and annealed for 6 h at $P_{res} \simeq 10$ Pa (*2*) and $P_{O_2} = 2 \times 10^4$ Pa (*3*). Annealing temperature 873 K [490].

between the maximum cathodic current and the degree of V_2O_5 reduction (concentration of vanadium(IV)) is illustrated in Fig. 6.7. A correlation between these quantities is observed. A pronounced kink on the curve in the region of 1.12% vanadium(IV) corresponds to the boundary of the homogeneity region of $V_2O_{5-\gamma}$ ($\gamma \leqslant 0.02$ according to [524]). This result confirms the validity of the hypothesis about a relationship between chemisorption and

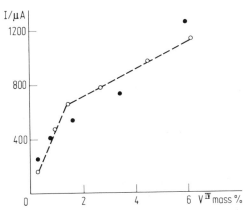

Figure 6.7. Current maximum ($E_p \approx 0.7$ V) versus concentration of V^{IV} centers in V_2O_5 partially reduced with sulphur at 620 K (filled circles) and in chemically pure V_2O_5 after annealing at 870 K (open circles) [490].

the concentration of anion vacancies (or reduced centers in the cation sublattice of the compounds), since V_2O_5 is characterized by oxygen non-stoichiometry [489], which brings about the appearance of quasi-free electrons and V^{4+} centers. The defect-formation equations can be written as

$$O_0^x \rightleftarrows 1/2O_2 + V_0^{\cdot\cdot} + 2e' \tag{6.11}$$

$$V_V^x + e' \rightleftarrows V_V' \tag{6.12}$$

where O_0^x is oxygen at its site; $V_0^{\cdot\cdot}$ is the oxygen vacancy; V_V^x (V_V') is the pentavalent (tetravalent) vanadium at the vanadium(V) site.

The loss of oxygen during the oxide reduction process is accompanied by ordering of the oxygen vacancies and the successive appearance of the V_4O_9 and V_6O_{13} oxides on the surface, with no significant rearrangement of the matrix structure. The concentration of oxygen vacancies becomes higher in the series V_2O_5–V_4O_9–V_6O_{13} and, correspondingly, an increase occurs in oxygen chemisorption and the signal of the second kind [490].

To elucidate the correlation of the signals of electrochemical transformations of chemisorbed oxygen and the defect structure of oxides, a study [525] was undertaken of the effect of iso- and heterovalent substitution into the structure of orthovanadates of magnesium and alkaline-earth metals (AEM), and of the thermodynamic conditions of synthesis. Among the advantages of the chosen systems is the possibility of obtaining extended solid solutions with different types of substitution and prevailing ionic conductivity [526]. This latter factor makes it possible to eliminate or minimize the effects of the water oxidation overvoltage, since the electron carriers are mainly responsible for the overvoltage in the abovementioned process [435, 505].

Thus, of the two factors affecting the magnitude of the oxygen cathodic current, the deciding one is the adsorption ability of the compound.

Isovalent Substitution. The anodic–cathodic–anodic curves of $Ca_3(V_{1-x}P_xO_4)_2$ solid solutions are depicted in Fig. 6.8. The cathodic voltammograms recorded beginning from the stationary potential or those following an anodic polarization are quite similar, differing only in the current magnitude. This shows a considerable contribution of chemisorbed oxygen reduction to the cathodic signals. The resolution of both the cathodic and anodic signals is impaired with increasing phosphorus concentration in the solid solution. The maximum current of chemisorbed oxygen reduction decreases as the phosphorus concentration in the sample rises. In the case of calcium orthophosphate the voltammogram practically coincides with that of the carbon-paste indifferent electrode. From these data it is evident that a decrease in the vanadium concentration in a calcium phosphate-vanadate solid solution is accompanied

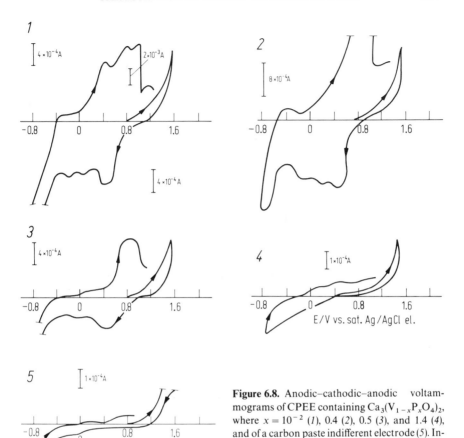

Figure 6.8. Anodic–cathodic–anodic voltammograms of CPEE containing $Ca_3(V_{1-x}P_xO_4)_2$, where $x = 10^{-2}$ (*1*), 0.4 (*2*), 0.5 (*3*), and 1.4 (*4*), and of a carbon paste indifferent electrode (*5*). Indifferent electrolyte 0.5 M H_2SO_4; $v = 0.08$ V/s.

by a decrease in the number of oxygen sorption centers which are represented by the V^{4+} ions [497].

An analysis of $Sr_{3(1-x)}Ba_{3x}(VO_4)_2$ solid solution voltammograms (Fig. 6.9), shows that the change in the sample composition has an effect on the magnitude of the signals and on the shape of the curves. The richest set, as regards the cathodic and related anodic signals, is recorded with the $Sr_{1.5}Ba_{1.5}(VO_4)_2$ sample.

With an increase of barium concentration, the cathodic currents decrease. This illustrates a decrease in oxygen adsorption on the surface of $Sr_{3(1-x)}$, $Ba_{3x}(VO_4)_2$ solid solutions with increasing x. As will be shown later, this is likely due to the lattice strengthening on transition from strontium to barium orthovanadate.

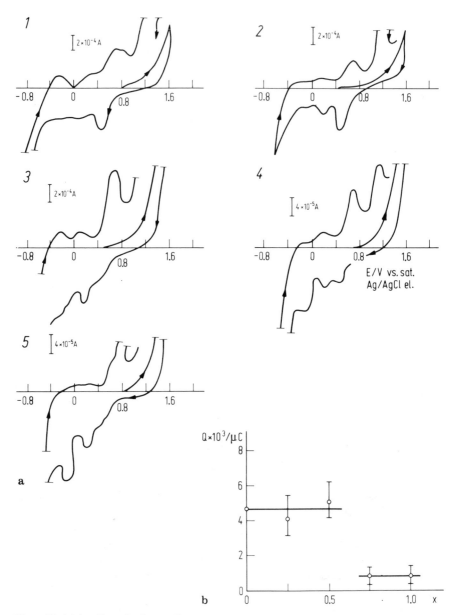

Figure 6.9. (a) Anodic–cathodic–anodic voltammograms of CPEE containing $Sr_{3(1-x)}$ $Ba_{3x}(VO_4)_2$ with $x = 0.0$ (1), 0.25 (2), 0.50 (3), 0.75 (4), and 1.00 (5). Indifferent electrolyte 0.5 M H_2SO_4; $v = 0.08$ V/s; $I_a = 0.625$ mA. (b) Dependence of electrical charge Q (equivalent to the amount of chemisorbed oxygen participating in the cathodic process) on the composition, x, of the solid solution.

Heterovalent Substitution. The AEM orthovanadates are essentially ionic crystal semiconductors. The vacancies in the cation (V_M'') and oxygen ($V_0^{\cdot\cdot}$) sublattices are the dominant defects. The introduction of donor and acceptor centers into the structure of orthovanadates results in a change of the concentration of existing defects, as well as in the appearance of new types of defects. Disordering of the orthovanadate lattice in the case of a heterovalent substitution can be represented by the following quasi-chemical expressions:

$$2AO_2 + 3MO \rightarrow 3M_M^x + 7O_0^x + 2A_V' + V_0^{\cdot\cdot} \tag{6.13}$$

$$2LnVO_4 \rightarrow 2Ln_M^{\cdot} + 2V_V^x + 8O_0^x + V_M'' \tag{6.14}$$

where M is an alkaline-earth metal, A is a fourth group element (Ti, Zr), and Ln is the lanthanide. An addition of Na as an acceptor into the structure of calcium orthovanadate results in predominant Na_{Ca}' and oxygen vacancies ($V_0^{\cdot\cdot}$) (over the investigated region of 0.1–0.3 mol% Na_3VO_4 added) [527].

The heterovalent substitution, in neither the AEM nor the vanadium sublattices, has any marked effect on the increase of signals, that is not accompanied by an increased capacity of orthovanadates for chemisorption of oxygen. Hence, the disordering that does not affect the concentration and the nature of electronic defects does not lead to an increase of oxygen sorption by the surface of the oxide, or of the number of the corresponding electrochemical responses.

Electronic Disorder. The data given in the previous section testify unambiguously to the absence of correlations between the concentration of oxygen vacancies arising from heterovalent substitution of the structural elements of complex oxides and the magnitude of the reduction currents of oxygen chemisorbed on the oxide surface. However, as follows from the papers [507, 521, 528], such correlations are observed. To find out the cause of this contradiction, the dependence of the oxygen chemisorption capacity on the vacancy concentration in the oxygen sublattice of complex oxide compounds produced by quenching of the samples or by their annealing at a lower partial oxygen pressure was examined.

The results obtained are given in Table 6.1. As can be seen, the magnitude of the first signal of chemisorbed oxygen, which was recorded after anodic polarization up to the preset current or up to the preset potential on quenched magnesium, calcium, strontium, and barium orthovanadates, or on calcium molybdate annealed at a lower partial oxygen pressure, increases compared to the samples that were subjected to slow cooling.

It also follows from Table 6.1 that quenching of the oxides from higher temperatures and annealing of these at a lower P_{O_2} does not significantly

Table 6.1. Influence of the Oxide Compound Synthesis Conditions on the Potential, E, Corresponding to $I = 0.3\,mA$, and on the Magnitude of the First Cathodic Maximum, I_m (after Polarization up to $E_a = 1.7\,V$)

Oxide	Synthesis Conditions	$E(V)$	$I_m(mA)$	$S_1(mA^a)$
$Mg_3(VO_4)_2$	Slow cooling	1.56	0.26	0.081
	Quenching from 1170 K	1.46	0.59	0.041
$Ca_3(VO_4)_2$	Slow cooling	1.05	0.55	0.032
	Quenching from 1370 K	1.10	1.10	0.261
$Sr_3(VO_4)_2$	Slow cooling	1.20	0.45	0.046
	Quenching from 1170 K	1.25	0.64	0.038
$Ba_3(VO_4)_2$	Slow cooling	1.34	0.030	0.025
	Quenching from 1370 K	1.38	0.23	0.028
$CaMoO_4$	Slow cooling	1.22^b	0.01^c	0.002
	Annealing at $P_{O_2} = 10^{-4}\,Pa$	1.24^b	0.02^c	0.009
	Annealing at $P_{O_2} = 10^{-14}\,Pa$	1.27^b	0.04^c	

[a] Standard deviation.
[b] Potential corresponding to $I_a = 0.1\,mA$.
[c] After anodic polarization up to 1.3 V.

influence the value of the water oxidation overvoltage. As a result, anodic polarization both at the preset current and at the preset potential leads to an increase in the cathodic currents of chemisorbed oxygen reduction. Similar results have been obtained from the study of nonstoichiometric $V_2O_{5-\gamma}$ [490].

The phenomena observed are due, most probably, to the processes that take place during thermal treatment of the samples studied. As is known [529], AEM orthovanadates are characterized by vacancy disorder. As the temperature is raised and the oxygen pressure is lowered, the structure of the compounds studied develops oxygen nonstoichiometry according to Eq. 6.11. The fact that the water oxidation overvoltage is almost independent of the orthovanadate thermal treatment conditions is probably due to localization of electrons arising from Eq. 6.11 on vanadium ions, thus forming V'_V centers which are oxygen sorption centers (Eq. 6.12).

The results attest to the fact that the maximum of the reduction current of oxygen chemisorbed on the oxide compound surface is unambiguously related to the oxygen sublattice disorder. An increase in the chemisorbed oxygen signal is observed only for such an ion disorder of the anion sublattice, which is accompanied by electronic disorder (Eq. 6.11).

Taking into account the results of the studies [507–520], we should acknowledge that these deductions also hold for signals of the first kind. The interpretation of the results obtained by the authors [507] from a study of doped titanium and zirconium oxides can hardly be accepted. In [525], an attempt was made to reproduce the signals [507] on ZrO_2, $Zr_{1-x}Sc_xO_{2-x/2}$, and $Ti_{1-x}Al_xO_{2-x/2}$ oxides using the techniques employed in [507, 525]. They failed to obtain signals of chemisorbed oxygen on these oxides, probably owing an increase in the activation energy of oxygen exchange with the gaseous phase and to a decrease in the chemisorbed oxygen amount with increasing strength of the compound crystal lattice. However, when studying titanium oxides synthesized in a reducing atmosphere whose diffractograms showed reflexes typical of oxides with lowest degrees of oxidation and EPR spectra of Ti^{3+} paramagnetic centers, the authors did reproduce the signals described in [507]. Note that, in that study, signals of larger magnitude were also registered on titanium dioxide samples calcined in air and at elevated temperatures than on the corresponding oxides calcined in oxygen atmosphere.

Then, oxygen sorption and oxygen reduction currents (signals of the first and second kinds) are determined by a disorder in the oxygen sublattice of the compound, accompanied by electronic disorder. The chemisorbed oxygen reduction currents can serve as a source of information only about such types of disorder in oxide compounds [435, 530, 531].

6.3.2. Energy Factors and Forms of Cathodic and Anodic Voltammograms

Figure 6.10 shows voltammograms of magnesium and AEM orthovanadates. Cathodic current maxima are observed in all the cases.

The magnitudes, number of signals, and potentials corresponding to the cathodic current maxima are dependent on the type of vanadate used. After a preliminary anodic polarization, similar cathodic signals, but markedly larger, were recorded. In the case of calcium and strontium vanadates, the signals are resolved better than in the case of $Mg_3(VO_4)_2$ and $Ba_3(VO_4)_2$.

Associated with the cathodic signals are the corresponding anodic signals (I′–III′) whose number and character depend, in addition to the type of vanadate used, on the potential at which the scan direction is changed. An increase of the cathodic polarization interval results in merging of signals II′ and III′. In this case phase transformations are liable to occur. The magnitude of chemisorbed oxygen reduction currents is largest for $Ca_3(VO_4)_2$, and regularly decreases in the $Ca_3(VO_4)_2$–$Sr_3(VO_4)_2$–$Ba_3(VO_4)_2$ series.

Since the signals are due largely to the electrochemical transformation of chemisorbed oxygen, these are also determined by (1) the ability of the compounds in the CPEE to adsorb oxygen and (2) thermodynamics and kinetics of the electron transfer. The oxidation reaction of a lower oxide into a higher one should be taken into account when studying the processes of electrochemical adsorption of oxygen on oxides [497]. However, no such

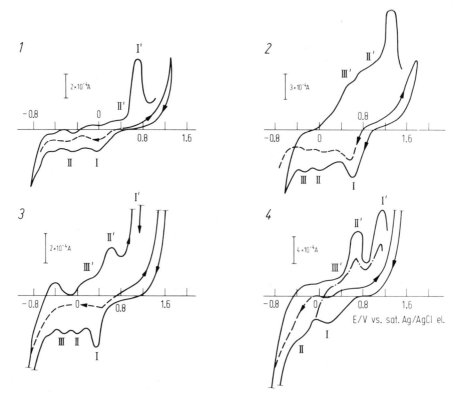

Figure 6.10. Anodic–cathodic–anodic (solid lines) and cathodic (dashed lines) voltammograms of $Mg_3(VO_4)_2$ (1), $Ca_3(VO_4)_2$ (2), $Sr_3(VO_4)_2$ (3), and $Ba_3(VO_4)_2$ (4). Indifferent electrolyte 0.5 M H_2SO_4. Scan rate (v) 0.08 V/s. The initial potential (E_i) was equal to the stationary potential (E_{st}). The linear potential scan was preserved during the interruption of the curves. Reference is the Ag/AgCl electrode in saturated KCl solution.

data are available for orthovanadates and therefore we will consider experimental results in connection with the crystalline and electronic structure and thermodynamics of the oxides mentioned above. Such an approach is justified, as follows from published data.

Boreskov [532] suggested a linear correlation between the exchange energy of the surface oxygen with the gaseous phase and the oxide formation enthalpy, which, in his opinion, should be regarded as a parameter characterizing the strength of the M–O bond. Winter [533] has questioned this idea and has established a correlation between the exchange reaction activation energy and the minimum O–O and M–O distances in isostructural oxide groups.

In the series of orthovanadates of alkaline-earth metals, which are closely related by their structure, an increase of their enthalpy of formation from binary oxides and of their melting temperatures with the cation ion radius [534] testifies to an enhancement of interatomic interaction.

Data on the energy parameters of the V–O bond can be obtained from optical measurements. The optical absorption of AEM orthovanadates in the region 3.5–4.0 eV is associated with the process of charge transfer from the oxygen environment to vanadium in the vanadium-oxygen tetrahedra [529]. The energy corresponding to the boundary of fundamental absorption of $M_3(VO_4)_2$ tends to increase with the growing atomic number of the AEM, except magnesium orthovanadate. This points to an increase in the binding energy of V–O. Taking into account these data, one expects a shift of the potential maximum of the major (the first) cathodic signal in the direction of more negative values in the series of calcium, strontium, and barium orthovanadates, since the signals are associated with the localization of negative charge by the chemisorbed oxygen, needed for building up the anion sites in the crystal structure and should depend on energy of structure formation itself. The described phenomenon has been substantiated by experimental data (Table 6.2).

The increase in the energy of the lattice and the V–O bond with an increase in the AEM radius is followed by lowering of the magnitudes of cathodic signals (Table 6.2). This points to a decrease in oxygen sorption on the surface. Hence, the signals can be used to evaluate the capacity of oxide compounds for oxygen sorption, which is of interest in predicting their catalytic activity. However, magnesium orthovanadate does not obey this rule which, according to Winter [533], is valid only for compounds having similar structures.

Table 6.2. Electrochemical and Thermodynamic Characteristics of $M_3(VO_4)_2$[a]

| Metal | Magnitude of the First Cathodic Signal (mA) After Anodic Polarization | | | | | | |
	Up to 0.43 mA	Up to 1.7 V	S_1	E_m(V)	T_{melt}(K) [534]	$-\Delta H^0$ (kJ/mol) [534]	$E_{f.abs.}$ (eV) [534]
Mg	0.10	0.26	0.08	0.40	1212	108.5	3.85
Ca	0.48	0.55	0.03	0.50	1380	321.7	3.50
Sr	0.27	0.45	0.04	0.45	1540	509.4	3.85
Ba	0.05	0.03	0.02	0.25	1690	696.4	4.00

[a]S_1—standard deviation of the magnitude of cathodic signal observed after anodic polarization up to a potential of 1.7 V; E_m—potential of the first cathodic maximum; T_{melt}—melting temperature; ΔH^0—enthalpy of orthovanadate formation from vanadium and AEM oxides; $E_{f.abs.}$—energy corresponding to the fundamental absorption boundary.

Indeed, strontium and barium orthovanadates are isostructural and crystallize to the rhombohedral structure (space group R32); calcium orthovanadate has a similar structure (space group R3c); magnesium orthovanadate exhibits however a different structure: it crystallize to the rhombic structure (space group Cmca) [535, p. 179].

An analysis of the crystal structure of the AEM orthovanadates has shown [529] that the presence of structurally nonequivalent oxygen is quite characteristic of such compounds. The number of nonequivalent positions occupied by oxygen ions tends to increase in the series from barium to calcium. Thereby the resolution of both the cathodic and anodic signals of chemisorbed oxygen is improved (see Fig. 6.10).

The relationship between the number of cathodic signals, their discreteness, and the number of nonequivalent positions of oxygen from the crystalchemical point of view is confirmed by an investigation of the homologous series of vanadium oxides: $V_2O_5-V_3O_7-V_4O_9-V_6-O_{13}-V_2O_4$ [490]. After anodic polarization of the CPEE containing these oxides, a spectrum of signals appears on the cathodic plot, whose magnitude is considerably higher than the electrochemical transformation signals proper. The magnitude, number and discreteness of the cathodic signals depend not only on the nature of the oxide but also on the value of the anodic polarization potential and on the acidity of the solution. Over the potential range 1.0 to -0.4 V only one maximum of cathodic current is registered at potentials 0.50 and 0.65 V for V_3O_7 and V_2O_4, respectively, while in the case of V_2O_5, V_4O_9, and V_6O_{13} several maxima are observed (Fig. 6.11). With the broadening of the amplitude of cathodic polarization of V_2O_5 up to -0.8 V, five signals are observed (Fig. 6.12). Associated with the cathodic signals are the corresponding anodic signals, their number and character also depend on the nature of the oxide and the potential at which the scan direction is changed.

In the $V_2O_5-V_2O_4$ series one finds a set of nonequivalent vanadium–oxygen bonds. Given below are the maximum differences in the bond lengths (Δl) in each vanadium oxide:

Oxide	V_2O_5 [536, 537]	V_3O_7 [538]	V_4O_9 [538]	V_3O_{13} [538]	V_2O_4 [536]
$\Delta l(\text{Å})$	1.20	0.50	1.40	0.64	0.30

The nonequivalence of the V–O bonds in the V_2O_5, V_4O_9, and V_6O_{13} oxides which exhibit a set of cathodic signals is much higher than in V_3O_7 and V_2O_4, where only one cathodic current signal is registered. Hence, the

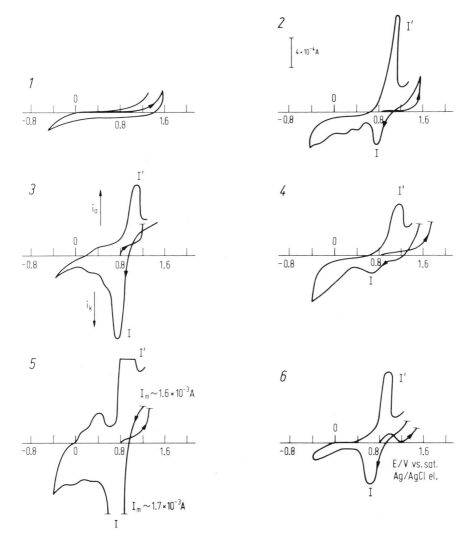

Figure 6.11. Anodic–cathodic–anodic voltammograms of CPEE (*1*) without vanadium oxide, (*2–6*) containing V_2O_5, V_4O_9, V_3O_7, V_6O_{13}, V_2O_4. Indifferent electrolyte 0.5 M H_2SO_4, $v = 0.02$ V/s, $E_i = 0.75$ V (*2*) and 0.80 V (*1, 3–6*). Anodic polarization up to the potential (E_a) 1.4 V (*2*) and 1.5 V (*1, 3–6*). The linear potential scan was preserved during the interruption of the curves.

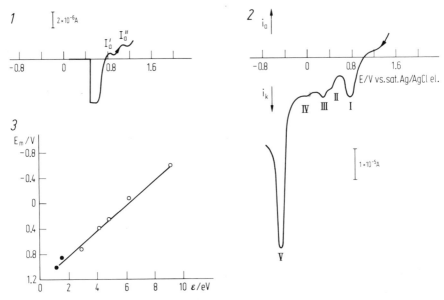

Figure 6.12. Anodic (*1*) and cathodic (*2*) voltammograms of CPEE containing V_2O_5. Indifferent electrolyte 0.5 M H_2SO_4. $E_i = 0.5$ V (*1*) and 1.5 V (*2*). Electrode was kept 15 s at $E = 1.5$ V. (*3*) Dependence of the peak potentials E_m on the energy values ε of maxima in the absorption spectra of V_2O_5 single crystals [539]. Peaks I'_a, I''_a (●) and I–V (○).

set of cathodic signals and their discreteness reflect the crystal-chemical non-equivalence of oxygen in the structure of the compounds under study.

An additional confirmation of the relationship between electrochemical transformations and the crystal and electronic structure can be found in the correlation between the energies of the maxima in the V_2O_5 optical absorption spectrum and the potentials of current maxima which are observed after preliminary anodic polarization of the samples (see Fig. 6.12). The optical absorption of V_2O_5 crystals is caused by the process of charge transfer from the oxygen environment to the vanadium ions [529], that is, this spectrum, like the electrochemical one, is determined by the electronic structure of the vanadium–oxygen groups.

An interpretation of the V_2O_5 optical absorption spectra is given [537] on the basis of quantum-mechanical calculations of the electronic structure of the VO_n^{5-2n} clusters. As a result, it is shown that the optical transitions in the region 2–3 eV can be explained by taking into account only the oxygen that has the weakest bond with the vanadium ion [537]. The absorption in the region can be correlated with the first cathodic maximum (signal I) having the lowest energy. Absorption processes in the region of higher energy values, correlating with the subsequent cathodic maxima, are well described in terms of the VO_n^{5-2n} cluster model, where $n = 4$ and 5. Below the boundary of self-absorption of V_2O_5 with $E \sim 2\,eV$, the absorption bands are due to $3d$-electrons of the tetravalent vanadium centers [539] which exhibit anodic current peaks during oxidation of partially electrochemically reduced VO_n^{5-2n} (peaks I'_a and I''_a in Fig. 6.12).

The appearance of anodic currents associated with the cathodic ones (see Figs. 6.10 and 6.11) points to the low stability of chemisorbed oxygen and to its localization on the surface of the solid in both its oxidized and reduced form. One can judge the degree of stability (reversibility) by considering the difference in potentials ($\Delta E_{1/2}$), where both the cathodic and anodic currents attain one half of their respective maximum magnitudes. The dependences of the cathodic current maxima and of the potential differences ($\Delta E_{1/2}$) on the composition of the samples change in an opposite way [508].

It follows that a correlation exists between the activation energy of exchange reactions of surface oxygen with the gaseous phase, the energy of the crystal lattice, the molar volume, and the M–O and O–O bond lengths in oxides, on the one hand, and the maximum reduction current of chemisorbed oxygen and the peak potential on the other hand. Such a correlation, in general terms, occurs because the energy and dimensional factors determine the electronic structure of a compound, which in turn has a substantial influence on the oxygen sorption processes [497].

The energy nonequivalence of oxygen sorbed at the surface during the anodic polarization process yields information on the crystal-chemical nonequivalence of oxygen in the oxide structure. The values of potentials corresponding to the cathodic current maxima are related to the change of the oxygen binding energy in the structure. The dependence of cathodic signals on the capacity of a solid substance for oxygen sorption can serve for predicting its catalytic activity.

6.4. DISCHARGE-IONIZATION OF HYDROGEN

Figure 6.13 shows cathodic–anodic voltammograms recorded during polarization of titanium dioxide and solid solutions with rutile structure. It can be seen that after cathodic polarization at a potential up to $-0.8\,V$ an anodic current is registered only in the case of the $Ti_{1-x}Nb_xO_2$ solid solutions. This current and also the respective amount of electric charge tend to increase with the increase in niobium oxide concentration up to $3\,mol\%$ [516] (see Table 6.3).

According to [540], disordering of the oxides under study can be re-

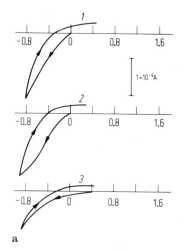

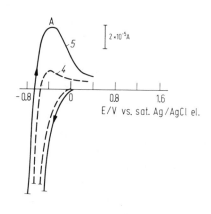

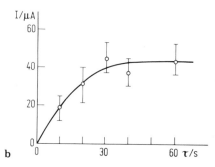

Figure 6.13. (*a*) Cathodic–anodic voltammo-grams recorded using a carbon paste indifferent electrode (*1*) and CPEE containing TiO_2 (*2*), TiO_2—0.05 mol % Al_2O_3 (*3*), TiO_2—1 mol % Nb_2O_5 (*4*), and TiO_2—5 mol % Nb_2O_5 (*5*). Indifferent electrolyte 1 M HCl; $v = 0.04$ V/s. (*b*) Dependence of hydrogen oxidation current on the cathodic polarization time at $I = 0.47$ mA of CPEE containing TiO_2—0.5 mol % Nb_2O_5.

Table 6.3. Hydroxonium-Ion Reduction Potentials, E, where the Cathodic Current Attains a Magnitude of 2×10^{-4} A, and Anodic Current, I_a, Corresponding to Oxidation of Hydrogen Adsorbed on the Surface of $TiO_2 + x$ mol % Nb_2O_5 [541]

x (mol % Nb_2O_5)	$-E$ (V)	I_a (µA)[a]	S_r[b]
0	0.94 ± 0.09	—	—
0.5	0.63 ± 0.02	8.62	0.47
1.0	0.50 ± 0.01	16.57	0.53
3.0	0.46 ± 0.06	47.92	0.40
5.0	0.47 ± 0.06	48.20	0.34
10.0	0.45 ± 0.06	49.10	0.37

[a] After cathodic potential scanning between 0.4 and -0.8 V.
[b] S_r—relative standard deviation.

presented by the quasi-chemical equations

$$Al_2O_3 \rightleftarrows 2Al'_{Ti} + 3O_0^x + V_0^{\cdot\cdot} \tag{6.15}$$

$$Nb_2O_5 \rightleftarrows 2Nb_{Ti}^{\cdot\cdot} + 4O_0^x + 1/2O_{2(g)} + 2e' \tag{6.16}$$

X-Ray analysis data suggest that in the $TiO_2-Nb_2O_5$ system in the concentration range below 14 mol% Nb_2O_5 samples quenched from a temperature of 1400 K are solid solutions with rutile structure. Returning to Table 6.3, it can be easily seen that at Nb_2O_5 concentrations not over 3 mol% the potentials, at which the cathodic current attains the preset magnitude, shift toward positive values, the anodic peak together with the corresponding electric charge increases. In the region of higher Nb_2O_5 concentrations, all of the quantities remain constant within the experimental error.

Two assumptions can be made as regards the causes of emergence of the cathodic and anodic currents: (i) reduction of niobium(V) ions and subsequent oxidation of Nb^{4+}; (ii) discharge-ionization of hydrogen according to the reaction

$$H_3O^+ + e' \rightleftarrows H_{ads} + H_2O \tag{6.17}$$

In both cases the amount of reduced substance can be expressed as a corresponding electrochemical charge that is proportional to the cathodic current flowing and to the electrolysis time. The current magnitude should be proportional to the concentration of the oxidized form of the substance [niobium(V) ions or H_3O^+].

The anodic peak should be naturally proportional to the amount of the substance reduced. The latter can either appear during electrochemical reduction or be present in the sample, for example, in the form of Nb^{4+} ions.

According to the first assumption, one should expect a direct dependence of the anodic peak on the niobium ion concentration in the solid solution. However, this is contradicted by the experiment.

The other assumption is less inconsistent. Indeed, the equilibrium of the reaction (Eq. 6.17) should shift noticeably to the right as the concentration of quasi-free electrons in the semiconductor rises. Therefore, the following should be observed: (i) a decrease in hydrogen overvoltage (the potential at which the cathodic current attains the preset value shifts toward positive values); (ii) an increase in the anodic current as the cathodic stage is extended till the surface is saturated with the adsorbed substance.

These phenomena are indeed observed experimentally.

The variation in the behavior of the "composition-property" dependence in the region of 3 mol% Nb_2O_5 and higher can be explained by the feature that, at these concentrations of the dopant, niobium ions localize electrons; that is, the system is transformed to the ternary TiO_2–Nb_2O_5–NbO_2 system. This is indirectly confirmed by the variations observed in the unit cell parameters and the refraction index [541], and by the growth of the ionic component of the total polarizability of the solid solutions.

In this case, an increase in the niobium ion concentration in the solid solution is not accompanied by a rise in the concentration of quasi-free electrons. All the parameters which depend on the concentration of quasi-free electrons (hydrogen overvoltage and anodic current) should not change, a situation which is in fact observed experimentally.

Thus, comparing the data given in Table 6.3 and in Fig. 6.13 one may conclude that the increase in the concentration of quasi-free electrons leads to a rise in the current of electrochemical oxidation of hydrogen formed during the cathodic polarization.

6.5. ELECTROCHEMICAL TRANSFORMATIONS OF SORPTION PRODUCTS AND THEIR RELATION TO BULK PROPERTIES OF SOLIDS

Although the abovementioned processes take place on the surface of solids, they are nonetheless determined, in principle, by the solids' bulk structure. In [507] it was shown that surface defects developing during attrition of poly-crystalline samples have no effect on the recorded signals whose magnitude depends on the concentration of the predominant bulk defects determined by the composition of the bulk phase. Shalaginov and co-workers [542] also attribute the effect to the "biographical" defects, that is, to those defects that develop during the formation of the substance. During attrition, defects stemming from mechanochemical effects are possible. However, the deciding defects are those developed during synthesis. This point is further confirmed in [495], where a symbatic change in the maximum of the chemisorbed oxygen reduction current and in the magnetic moment with the composition of vanadium oxide–sodium bronze, type β, was observed. Using single crystals of the bronze, a richer signal spectrum was registered [509] than that obtained on polycrystalline samples in powder form.

The extrema of bulk characteristics (density, specific magnetization, unit cell parameters) of $SrO \cdot nFe_2O_3$ and minimum values of chemisorbed oxygen reduction current versus composition are found for stoichiometric strontium hexaferrite [543].

Figure 6.14 displays results obtained from a study of magnesium, strontium, and barium orthovanadates which were held at the quenching temperature and then rapidly cooled. X-Ray analysis of these orthovanadates has shown that, although no new lines were registered in the diffractograms, the unit cell

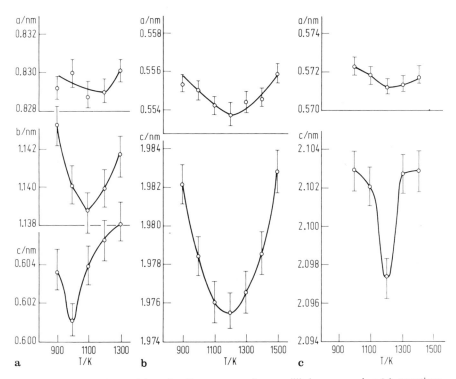

Figure 6.14. Dependence of the unit cell parameters of nonequilibrium magnesium (*a*), strontium (*b*), and barium (*c*) orthovanadates on the quenching temperature [546].

parameters have a minimum as a function of the quenching temperature [544].

Figure 6.15 shows the dependence of the amount of chemisorbed oxygen on the quenching temperature of the sample. Since the peaks are rather poorly resolved, the amount of chemisorbed oxygen is expressed as the ratio of the electric charge flowing off cathodically up to a potential of -0.4 V after the anodic polarization of the quenched samples to the electric charge flowing off cathodically after the anodic polarization of the respective equilibrium vanadates. Here it is unessential that in the case of barium orthovanadate the ascending branch only of the maximum could be measured. It can be seen that the ability of samples to sorb oxygen as well as the unit cell parameters have an extremum that depends on the quenching temperature.

Within the range of quenching temperatures corresponding to the extrema, the character of anodic–cathodic voltammograms changes (see Fig. 6.16); that is, resolution of the signals is improved, and a cathodic current appears at

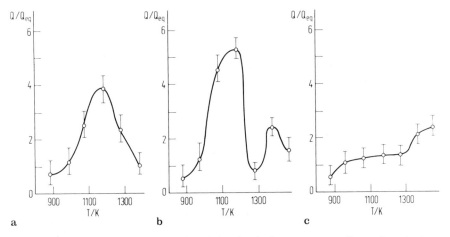

Figure 6.15. Dependence of Q/Q_{eq} (Q and Q_{eq} being electric charges corresponding to the reduction of oxygen chemisorbed on nonequilibrium and equilibrium samples) of magnesium (a), strontium (b), and barium (c) orthovanadates during their anodic polarization on the quenching temperature. Indifferent electrolyte 0.5 M H_2SO_4.

more positive potentials. This effect is particularly pronounced in the case of barium orthovanadate which yields a new cathodic signal registered at 0.6 V.

The experimental results may be interpreted by taking into account the changes in stoichiometry, with regard to the oxygen, in the vanadates studied. It is known [536] that the concentration of vacancies in the oxygen sublattice ($V_O^{\cdot\cdot}$) of oxides tends to increase as temperature is raised. The exchange reaction with oxygen is described by Eq. 6.11. Electrons can be localized on the vanadium ions (Eq. 6.12). These sorption centers (V_V') increase the oxygen sorption capacity of the oxide. Quenching of the samples "freezes" the thermal defects of the oxygen sublattice, as a result of which an increased amount of oxygen chemisorbed on the sample surface should be registered. This has been actually observed on samples quenched from temperatures of $\leqslant 1170$ K in the case of Mg and Sr, and $\leqslant 1400$ K in the case of Ba (see Fig. 6.15).

The change in the spectrum of oxygen chemisorbed on vanadates that have been quenched from temperatures above 1170 K may serve as evidence of the change in oxygen disorder. The change of the orthovanadate lattice parameters (see Fig. 6.14) points to the bulk nature of this process. The increase in the unit cell parameters of $Sr_3(VO_4)_2$ quenched from temperatures above 1170 K should be related, according to the paper [544], to the possible transfer of strontium cations to interstitial sites.

Defects as proposed by Wadsley [545] are more likely to form. Wadsley

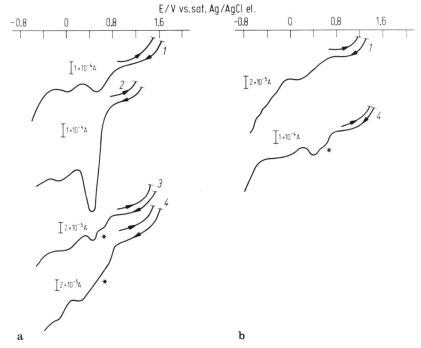

Figure 6.16. Anodic–cathodic voltammograms of CPEE containing strontium (*a*) and barium (*b*) orthovanadates. Quenching temperature: 870 (*1*), 1170 (*2*). 1270 (*3*), and 1470 K (*4*). Indifferent electrolyte 0.5 M H_2SO_4.

supposes that, at some critical concentration of oxygen vacancies, oxygen polyhedra join together along their edges, and thus crystallographic shift planes are formed. In this case, no change of the cation coordination is observed with a decrease in the effective concentration of oxygen vacancies.

In the orthovanadates considered here, such defects can be centers composed of two vanadium-oxygen complexes having a common edge. In such centers, as well as in the lowest vanadium oxides, vanadium is closely arranged, and two electrons participate in the $V^{4+}-V^{4+}$ bond. These centers are not paramagnetic and fail to yield EPR signals. Formation of such defects should undoubtedly influence the vanadium-oxygen environment and have an effect on the character of the polarization curves (see Fig. 6.16). The oxygen sorption capacity will decrease as a result of a drop in the concentration of the available sorption centers. A study of infrared spectra of nonequilibrium strontium orthovanadate has confirmed that in the case of samples quenched from temperatures higher than 1170 K an additional absorption band with a maximum at approximately $730\,cm^{-1}$ has been registered in the region of vanadium-oxygen charge transfer. A similar band appears also in the infrared spectra of strontium vanadate annealed at a lower oxygen partial pressure.

The defect mechanism proposed by Wadsley is not in conflict with the observed increase in the lattice parameters of quenched vanadates. In fact, the latter trend can be due to an increase in the electrostatic repulsion of non-compensated localized charges of the cations originating in places where oxygen left its sublattice. A quite similar phenomenon has been observed, for example, in the case of zirconium dioxide [546].

Figure 6.16 depicts a signal (∗) registered in the region of about $0.6\,V$ on the cathodic plot of the polarization curves of strontium and barium ortho-vanadates quenched from temperatures of 1270 and 1470 K, respectively. This suggests participation of weakly bound oxygen in the process [490]. In the case of $Ba_3(VO_4)_2$, this effect shows up for samples quenched from higher temperatures. Barium orthovanadate is characterized by the highest crystal lattice energy compared to other AEM orthovanadates, Table 6.2 [534], and hence in this case one expects the emergence of oxygen sorption centers at sufficiently high temperatures.

REFERENCES

1. Zolotov, Yu. A. (1977), *Essays on Analytical Chemistry*, Khimia, Moscow.

2. Orient, I. (1981), *Zh. Anal. Khim.*, **36**: 578.

3. Kazarinov, V. E. (Ed.) (1981), *Double Layer and Electrode Kinetics*, Nauka, Moscow.

4. Kissinger, P. T., Heineman, W. R. (Eds.) (1984), *Laboratory Techniques in Electroanalytical Chemistry*, Marcel Dekker, New York.

5. Bond, A. M. (1980), *Modern Polarographic Methods in Analytical Chemistry*, Marcel Dekker, New York.

6. Bond, A. M., Luscombe, D. L., Tan, S. N., Walter, F. L. (1990), *Electroanalysis*, **2**: 195.

7. Hernandez-Brito, J. J., Cardona Castellano, P., Perez-Pena, J., Gelado-Cabellero, M. D. (1990), *Electroanalysis*, **2**: 401.

8. Bersier, P. M., Bersier, J. (1985), *CKC Crit. Rev. Chem.*, **16**: 15, 65.

9. Bersier, P. M., Bersier, J. (1985), *CKC Crit. Rev. Chem.*, **16**: 81.

10. Orient, I. M., Markusova, B. A. (1972), *Electrochemical Methods of Materials Analysis*, p. 95, Metallurgiya, Moscow.

11. Stromberg, A. G., Orient, I. M., Kameneva, T. M. (1982), *Zh. Anal. Khim.*, **37**: 2245

12. Jahnson, D. C., Ryan, M. D., Wilson, G. S. (1984), *Anal. Chem.*, **56**: 7R.

13. Ryan, M. D., Wilson, G. S. (1982), *Anal. Chem.*, **54**: 20R.

14. Brainina, Kh. Z., Neiman, E. Ya. (1982), *Solid Phase Reactions in Electroanalytical Chemistry*, Khimia, Moscow.

15. Vydra, F., Stulik, K., Juláková, E. (1976), *Electrochemical Stripping Analysis*, Ellis Harwood, Chichester.

16. Kaplan, B. Ya. (1978), *Pulse Polarography*, Khimia, Moscow.

17. Heineman, W., Kissinger, P. (1980), *Anal. Chem.*, **52**: 139R.

18. Kalvoda, R. (1990), *Electroanalysis*, **2**: 341.

19. Smyth, M. R., Vos, J. G. (Eds.) (1991), *Analytical Voltammetry*, Elsevier, Amsterdam.

20. Florence, T. M. (1983), *Anal. Proc.*, **20**: 552.

21. Brainina, Kh. Z., Vidrevich, M. B. (1981), *J. Electroanal. Chem.*, **121**: 1.

22. Brainina, Kh. Z., Vidrevich, M. B. (1985), *Zavodsk. Lab.*, **51**: 3.

23. Kopanica, M., Opekar, F. (1987), in Kalvoda, R. (Ed.), *Electroanalytical Methods in Chemical and Environmental Analysis*, Plenum, New York.

24. Wang, J. (1989), in Bard, A. J. (Ed.), *Electroanalytical Chemistry*, Vol 16, Marcel Dekker, New York.

25. Wang, J. (1990), *Fr. Z. Anal. Chem.*, **337**: 508.

26. Barker, G. C., Jenkins, I. L. (1952), *Analyst*, **77**: 685.

27. Bersier, P. M. (1987), *Anal. Proc.*, **24**: 44.

28. Brainina, Kh. Z., Neiman, E. Ya., Slepushkin, V. V. (1988), *Inverse Electroanalytical Methods*, Khimia, Moscow.

29. Wang, J. (1985), *Stripping Analysis: Principles, Instrumentation and Applications*, VCH Publishers, Deerfield Beach.

30. Brainina, Kh. Z. (1981), *Bioelectrochem. Bioenerg.*, **8**: 479.

31. Boubliková, P., Vojtisková, M., Paleček, E. (1987), *Anal. Lett.*, **20**(2): 275.

32. Brainina, Kh. Z., Hodos, M. Ya. (1988), in *Wissenschaftliche Tagung der Technischen Universität*, Karl-Marx-Stadt (N10).

33. Brainina, Kh. Z. (1982), *Z. Anal. Chem.*, **312**: 428.

34. Brainina, Kh. Z., Tchernyshova, A. V., Stozhko, N. Yu., Kalnyshevskaya, L. N. (1989), *Analyst*, **114**: 73.

35. Göpel, W., Messe, J., Zemel, J. N. (Eds.) (1991), *Sensors: A Comprehensive Survey*, Vols. 2–3, VCH, Weinheim.

36. Bond, A. M., Heritage, J. D., Thorman, W. (1986), *Anal. Chem.*, **58**: 1063.

37. Brainina, Kh. Z. (1987), *Talanta*, **34**: 41.

38. Brainina, Kh. Z. (1972), *Inverse Voltammetry of Solid Phases*, Chimiya, Moscow. Brainina, Kh. Z. (1974), *Stripping Voltammetry in Chemical Analysis*, Halsted, New York.

39. Kolb, D. M., Przasnyski, M., Gerischer, H. (1974), *J. Electroanal. Chem.*, **54**: 25.

40. Schmidt, E., Gugax, H. R. (1966), *J. Electroanal. Chem.*, **12**: 300.

41. Brainina, Kh. Z., Roitman, L. I., Kalnishevskaya, L. N., and Podkorytov, E. M. (1980), *J. Electroanal Chem.*, **106**: 235.

42. Brainina, Kh. Z., Kaisin, A. V. (1983), *Elektrokhim.*, **19**: 87.

43. Fletcher, S., Halliday, C. S., Gates, D., Westcott, M., Zwin, T., Nelson, G. (1983), *J. Electroanal. Chem.*, **159**: 267.

44. Polukarov, Yu. M. (1985), *Physical Chemistry in Modern Problems*, Khimia, Moscow.

45. Montagu-Pollock, H. M., Rhodin, T. N., Souihon, M. J. (1968), *Surf. Science*, **12**: 1.

46. Suru, R., Thakour, A. P., Chopra, K. L. (1975), *J. Appl. Phys.*, **46**: 2574.

47. Scharifker, B. R., Mostany, J. (1984), *J. Electroanal. Chem.*, **177**: 13.

48. Brainina, Kh. Z., Shishkin, G. I., Roitman, L. I., Kalnishevskaya, L. N. (1978), *Zh. Anal. Khim.*, **33**: 1708.

49. Trassatti, S. (1977), *Advances in Electrochemistry and Electrochemical Engineering*, **10**: 213.

50. Markov, I., Kashchiev, D. (1972), *J. Cryst. Growth*, **16**: 170.

51. Fletcher, S. (1983), *Electrochim. Acta*, **28**: 917.

52. Mostany, J., Mozota, J., Scharifker, B. R. (1984), *J. Electroanal. Chem.*, **177**: 25.

53. Mayer, C., Jüttner, K. (1982), *Electrochim. Acta*, **27**: 1609.

54. Lava, S., Prasada, R. T., Prabhakara, R. G. (1986), *Electrochim. Acta*, **31**: 343.

55. Brainina, Kh. Z., Nikiforov, V. V. (1989), *Electrokhim.* **25**: 1237.

56. Krebbs, W. M., Roe, D. K. (1967), *J. Electrochem. Soc.*, **114**: 892.

57. Mamaev, A. I. (1985), in *Electrochemical Methods of Analysis*, Proceedings II, All-Union Conference on Electrochemical Methods of Analysis, June 4–6, 1985, TPI, Tomsk.

58. Kaisin, A. V., Shishkin, G. I., Beletskaya, A. A. (1984), in *Development and Applications of Polarography and Cognate Methods*, Proceedings VIII, All-Union Meeting on Polarography, Dnepropetrovsk, April 26–28, 1984, Part 1, DKhTI, Dnepropetrovsk.

59. Markov, I., Stoycheva, E., Dobrev, D. (1978), in *Communications of the Department of Chemistry of Bulgarian Academy of Science*, **11**: 377.

60. Vitanov, T., Sevastyanov, E., Stoinov, Z., Budevski, E. (1969), *Elektrokhim.*, **5**: 238.

61. Onischenko, A. V., Malov, Yu. I. (1982), *Fiz. Met. Metalloved.*, **54**: 94.

62. Malov, Yu. T., Korolkov, V. A., Markov, A. A. (1977), *Elektrokhim.*, **13**: 1243.

63. Korolkov, V. A., Malov, Yu. I., Markov, A. A. (1976), *Elektrokhim.*, **12**: 595.

64. Scholz, F., Nitschke, L., Henrion, G. (1988), *Zh. Anal. Khim.*, **18**: 1166.

65. Scholz, F., Nitschke, L., Henrion, G. (1987), *Anal. Chim. Acta*, **186**: 39.

66. Scholz, F., Nitschke, L., Henrion, G. (1987), *Z. Chem.*, **27**.

67. Scholz, F., Nitschke, L., Henrion, G. (1987), *J. Electroanal. Chem.*, **224**: 303.

68. Sinyakova, S. I., Markova, I. V., Galfayan, N. G. (1965), *Komis. Anal. Khim. AN SSSR*, **15**: 164.

69. Florence, T. M. (1980), *Anal. Chim. Acta*, **119**: 217.

70. Batley, G. E., Florence, T. M. (1974), *J. Electroanal. Chem.*, **55**: 23.

71. Bond, A. M. (1975), *Anal. Chim. Acta*, **74**: 163.

72. Brook, B. S. (1972), *Polarographic Methods,* Energiya, Moscow.

73. Doronin, A. N., Kabanova, O. L. (1965), *Zh. Anal. Khim.*, **20**: 1321.

74. Jagner, D., Graneli, A. (1976), *Anal. Chim. Acta*, **83**: 19.

75. Lamberts, L. (1981), *Anal. Chim. Acta*, **132**: 23.

76. Anderson, L., Jagner, D., Josefson, M. (1982), *Anal. Chem.*, **54**: 131.

77. Florence, T. M. (1970), *J. Electroanal. Chem.*, **27**: 273.

78. Zakharchuk, N. F., Valisheva, N. A., Yudelevich, I. G. (1980), *Izv. SO AN SSSR, Ser. Khim.*, **N9**/4: 33.

79. Demin, V. A. (1985), *Voltammetry of Organic and Inorganic Compounds*, p. 201, Nauka, Moscow.

80. Brainina, Kh. Z. (1980), *Elektrokhim.*, **16**: 678.

81. Brainina, Kh. Z., Tchernysheva, A. V., Stozhko, N. Yu. (1980). *Elektrokhim.*, **16**: 1874.

82. Brainina, Kh. Z., Kalnishevskaya, L. N. (1987), *Elektrokhim*, **23**: 222.

83. Roizenblat, E. M., Krapivkina, T. A. Veretina, G. N. (1974), *Zavodsk. Lab.*, **40**: 370.

84. Kritsotakis, K., Tobshall, H. G. (1985), *Fr. Z. Anal. Chem.*, **320**: 156.

85. Dominova, I. G., Kolpakova, N. A., Stromberg, A. G. (1977), *Zh. Anal. Khim.*, **32**: 1980.

86. Chulkina, L. S., Sinyakova, S. I., Zarinski, V. A. (1973), *Zh. Anal. Khim.*, **28**: 1221.

87. Kolpakova, N. A., Nemova, V. V., Stromberg, A. G. (1971), *Zh. Anal. Khim.*, **26**: 1217.

88. Shifris, B. S., Rakhmanberdyyeva, A. D., Bolshakova, N. A., Nazarov, B. F., Stromberg, A. G. (1977), VINITI, 8 June 1975, TPI, Tomsk.

89. Popov, G. N., Pnyov, V. V., Zakharov, M. S. (1972), *Zh. Anal. Khim.*, **17**: 2456.

90. Krapivkina, E. M., Roizenblat, E. M., Nosacheva, V. V., Zaretski, L. S., Utenko, V. S. (1974), *Zh. Anal. Khim.*, **29**: 1818.

91. Lexa, J., Stulik, K. (1983), *Talanta*, **30**: 845.

92. Kamenev, A. I., Rodiontsev, I. A., Agasyan, P. K. (1984), *Zh. Anal. Khim.*, **39**: 2142.

93. Kamenev, A. I., Mustafa, I., Agasyan, P. K. (1985), *Zh. Anal. Khim.*, **40**: 1483.

94. Kamenev, A. I., Mustafa, I., Agasyan, P. K. (1984), *Zh. Anal. Khim.*, **39**: 1242.

95. Kaplin, A. A., Portnyagina, E. O. (1981), *Zh. Anal. Khim.*, **36**: 1965.

96. Rosey, R., Andrews, R. (1981), *Anal. Chim. Acta*, **124**: 107.

97. Neiman, E. Ya., Ponomarenko, G. B. (1975), *Zh. Anal. Khim.*, **30**: 1132.

98. Kostromina, E. I. (1981), In Proc. Conf. of Young Scientists Dedicated to 225 Anniversary of MGU, Part II, Moscow.

99. Krapivkina, T. A., Roisenblat, E. M., Kalambet, G. A., Nosacheva, V. V. (1977), *Zh. Anal. Khim.*, **32**: 293.

100. Kiekens, P. et al. (1984), *Analyst*, **109**: 909.

101. Kiekens, P., Semmerman, E., Verbeek, F. (1984), *Talanta*, **31**: 693.

102. Roizenblat, E. M., Veretina, G. N. (1974), *Zh. Anal. Khim.*, **29**: 2376.

103. Wang, J., Turner, K. J. (1985), *Anal. Chem.*, **58**: 1086.

104. Neiman, E. Ya., Sumenkova, M. F., Nemodruk, A. A. (1975), *Anal. Chem.*, **30**: 1012.

105. Neiman, E. Ya. (1980), *Anal. Chim. Acta*, **113**: 277.

106. Valisheva, N. A., Zakharchuk, N. F., Yudelevich, I. G. (1980), *Izv. SO AN SSSR, Ser. Khim.*, **N12/5**: 88.

107. Brainina, Kh. Z., Kalnis˙ evskaya, L. N. Vilchinskaya, E. A., Khanina, R. M., Forshtadt, V. M. (1989), in Proceedings of the Third Beijing Conference and Exhibit on Instrum. Analysis, 27–30 October 1989, Beijing, China, F71.

108. Pratt, K. W., Koch, W. (1988), *Anal. Chim. Acta*, **215**: 21.

109. Barikov, V. G., Rozhdestvenskaya, Z. B., Songina, O. A. (1969), *Zavodsk. Lab.*, **35**: 776.

110. Brainina, Kh. Z., Vidrevich, M. B. (1985), *Zavodsk. Lab.*, **51**: 3.

111. Plambeck, J. (1985), *Electrochemical Methods of Analysis: Theory and Applications*, Mir, Moscow.

112. Tenygl, I. (1982), *Zavodsk. Lab.*, **48**: 4.

113. Jager, E., Molla, J. A. Gupta, S. (1983), In Proc. Workshop Electrochem. Carbon, Cleveland, Ohio, 17–19 Aug. 1983, Pennington, N. Y.

114. Thomas, J. (1983), *Anal. Proc.*, **20**: 565.

115. Andruzzi, R., Trazza, A. (1981), *Talanta*, **28**: 839.

116. Posly, R. S., Andrews, K. W. (1980), *Anal. Chim. Acta*, **119**: 55.

117. Wang, J., Greene, B. (1983), *J. Electroanal. Chem.*, **154**: 261.

118. Chaunguo, Cui (1984), *Talanta*, **31**: 221.

119. Wan, Z. (1985), *Environ. Chem.*, **4**: 51.

120. Karnaukhov, A. I. (1985), *Ukr. Khim. Zhurn.*, **51**: 1049.

121. Zebreva, A. I., Levitskaya, S. A., Aldamzhairova, S. Kh., Polokovnikova, L. S. (1983), *Izv. Vuzov. Khim. i khim. tekhnol.*, **26**: 715.

122. Bruckenstein, S., Mack, R. (1984), *Anal. Chim. Acta*, **158**: 1.

123. Kletenik, Yu. B., Zakharova, N. M., Bek, R. Yu. Zamyanin, A. P. (1985), *Izv. SO AN SSSR, Ser. Khim.* 2/1: 93.

124. Khanina, R. M., Tataurov, V. P., Brainina, Kh. Z. (1988), *Zavodsk. Lab.*, **54**(2): 1.

125. Kabanova, O. A., Goncharov, Yu. A. (1976), *Zh. Anal. Khim.*, **31**: 902.

126. Roizenblat, E. M. (1983), *Methods of Analyzing Materials Used Electronics*, Nauka, Moscow.

127. Mentus, Z., Mentus, S., Marinković, N., Lausević (1990), *J. Electroanal. Chem.*, **283**: 449.

128. Rice, R. J., Pontikos, N. M., McCreery, R. L. (1990), Abstract No. 308, Pittsburgh Conference and Exposition on Analytical Chem. and Applied Spectroscopy, 5–9 March 1990, New York.

129. Wang, J., Martiner, T., Yaniv, D. R., McCormick, L. (1990), *J. Electroanal. Chem.*, **286**: 265.

130. Wang, J. (1981), *Anal. Chem.*, **52**: 2280.

131. Edmonds, T. E. (1985), *Anal. Chim. Acta.*, **175**: 1.

132. Henriques, H. P., Fogg, A. G. (1985), *Analyst*, **110**: 79.

133. Lexa, J. (1985), *Talanta*, **32**: 1027.

134. Moskvitina, O. A., Salikhdzhanova, R. M. F. (1980), *Modern Methods of Chemico-Analytical Monitoring*, p. 75, Nauka, Moscow.

135. Svoboda, G. J., Sottery, J. P., Anreson, C. W. (1984), *Anal. Chim. Acta*, **166**: 297.

136. Kaplin, A. A., Pichugina, V. M., Svischenko, N. M. (1985). *Zavodsk. Lab.*, **51**: 14.

137. Baranski, A. S., Quan, H. (1986), *Anal. Chem.*, **58**: 407.

138. Colyer, C. L., Luscombe, D., Oldham, K. B. (1990), *J. Electroanal. Chem.*, **283**: 379.

139. Gruzkova, N. A. (1982), *Zavodsk. Lab.*, **48**: 12.

140. Wikiel, K., Kublic, Z. (1984), *J. Electroanal. Chem.*, **161**: 269.
141. Gunasingham, H., Ang, K. P., Ngo, C. C. (1985), *J. Electroanal. Chem.*, **186**: 51.
142. Kiekens, P., Van den Broecke, F., Bogaert, M., Temmerman, E. (1983), *Bull. Soc. Chim. Belg.*, **92**: 929.
143. Kramer, C. J. M., Guo-Kui, Yu., Duinker, J. C. (1984), *Anal. Chim. Acta*, **164**: 163.
144. Wang, J., Hutshihs-Kumar, L. D. (1986), *Anal. Chem.*, **58**: 402.
145. Schulze, G., Frensel, W. (1984), *Anal. Chim. Acta*, **159**: 95.
146. Jennings, V. J., Morgan, J. E. (1985), *Analyst*, **110**: 121.
147. Ciczkowska, M., Stojek, Z. (1985), *J. Electroanal. Chem.*, **191**: 101.
148. Gustavsson, J., Lundström, K. (1983), *Talanta*, **30**: 959.
149. Lundström, K. (1983), *Anal. Chim. Acta*, **146**: 97.
150. Gustavsson, J., Lundström, K. (1984), *Fr. Z. Anal. Chem.*, **317**: 388.
151. Kauffmann, J. M., Landet, A., Patriarche, G. J., Christian, G. D. (1982), *Talanta*, **29**: 1072.
152. Kauffmann, J. M., Montener, Th., Van den Baeck, J. L., Patriarche, G. J. (1984), *Microchim. Acta*, **1**: 95.
153. Kauffmann, J. M., Landet, A., Patriarche, G. J. (1982), *Anal. Chim. Acta*, **135**: 153.
154. Kauffmann, J. M., Landet, A., Patriarche, G. J. (1982), *Anal. Lett.*, **A15**: 763.
155. Henriquess, H. P., Fogg, A. G. (1984), *Analyst*, **109**: 1195.
156. Bull, R. A., Fan Fu-Ren, F., Bard, A. (1984), *J. Electrochem. Soc.*, **129**: 1009.
157. Stewart, E. E., Smart, R. B. (1984), *Anal. Chem.*, **56**: 1131.
158. Wang, J., Hutchins, L. D. (1985), *Anal. Chem.*, **57**: 1536.
159. Murray, R. W. (1984), in Bard, A. I. (Ed.), *Chemically Modified Electrodes in Electroanalytical Chemistry*, Vol. 13, p. 191, Marcel Dekker, New York.
160. Labuda, J. (1990), *Zh. Anal. Khim.*, **45**: 629.
161. Van der Linden, W. E., Dekker, J.'W. (1980), *Anal. Chim. Acta*, **119**: 1.
162. Berfand de Vismes, Bedioni Fethi, Devynck, J., Bied-Charreton, C. (1985), *J. Electroanal. Chem.*, **187**: 197.
163. Ikariyama, Y., Heinemann, W. R. (1986), *Anal. Chem.*, **58**: 1803.
164. Beck, F., Hülser, P. (1990), *J. Electroanal. Chem.*, **280**: 159.
165. Grinnshaw, J., Perera, S. D. (1990), *J. Electroanal. Chem.*, **278**: 279.
166. John, R., Wallace, G. G. (1990), *J. Electroanal. Chem.*, **283**: 87.
167. Ikariyama, Y., Galiadsatos, C., Heineman, W. R., Iamanchi, S. (1987), *Sens Actuators*, **12**: 455.
168. Wang, J., Lu, Z. (1989), *Electroanal. Chem.*, **266**: 287.
169. Hernandez, L., Melguizo, J. M., Blanko, M. H., Hernandez, P. (1989), *Analyst*, **114**: 397.
170. Kuaizhi, L., Hailan, L., Qingguo, W. U. (1989), Proceedings of the Third Beijing Conference and Exhibit on Instrum. Analysis, 27–30 October 1989, Beijing, China.
171. Matuszewski, W., Trojanowicz, M. (1988), *Analyst*, **113**: 735.

172. Tuzhi, P., Qingzchao, S. (1989), Proc. 3rd Beijing Conf. and Exhib. on Instrum. Analysis, 27–30 October 1989, China, F 97.

173. Ravichandram, K., Baldwin, R. P. (1983), *Anal. Chem.*, **55**: 1586.

174. Trojanowicz, M., Matuszewski, W. (1989), *Talanta*, **36**: 680.

175. Lexa, J. (1989), *Talanta*, **36**: 843.

176. Zak, J., Kuwana, Th. J. (1983), *J. Electroanal. Chem.*, **104**: 5514.

177. Zak, J., Kuwana, Th. J. (1982), *J. Am. Chem. Soc.*, **104**: 4736.

178. Boyer, A., Kolcher, K., Pietsch, R, (1990), *Electroanalysis*, **2**: 155.

179. Xue, Z. L., Karagölzler, A. E., Ataman, O. Y., Galal, A., Amer, A., Shabana, R., Zimmer, H., Mark, H. B. (1990), *Electroanalysis*, **2**: 1.

180. Garder-Torresdey, I., Darnall, D., Wang, J. (1988), *Anal. Chem.*, **60**: 72.

181. Wang, J., Benakdar, M. (1988), *Talanta*, **35**: 277.

182. Tanaka, S., Yoshida, H. (1989), *Talanta*, **36**: 1044.

183. Dong, S., Wang, Y. (1988), *Talanta*, **35**: 818.

184. Budnikov, G. K., Medyantseva, E. P., Volkov, A. V., Aronzon, S. S. (1983), *Zh. Anal. Khim.*, **38**: 1283.

185. Tomoo, Miwa, Li-Tong, Jin (1984), *Anal. Chim. Acta*, **160**: 135.

186. Van den Berg C. M. G. (1984), *Anal. Lett.*, **A17**: 2141.

187. Gomathi, H., Rao, G., Prabhakara M. (1986), *J. Electroanal. Chem.*, **190**: 85.

188. Kalvoda, R. (1979), In *Euroanalysis 3 Rev. Analyt. Chem.*, Pergamon, London.

189. Lubert, R. H., Schnurrbusch, M., Thomas, A. (1982), *Anal. Chim. Acta*, **144**: 123.

190. Nikitina, N. A., Tchernysheva, A. V., Stozhko, N. Yu., Brainina, Kh. Z. (1990), In *Analysis-90*. Proceedings of the All Union Conference on Modern Methods of Analysis of Metals, Alloys and Environmental Materials, Part 1, 1990, Izhevsk.

191. Schnurrbusch, M., Lubert, K-H., Thomas, A. (1983), *Z. Chem.*, **23**: 194.

192. Price, J. F., Baldwin, R. P. (1980), *Anal. Chem.*, **52**: 1940.

193. Downard, A. J., Ripton, H., Powell, J., Shuanghua, X. (1991), *Anal. Chim. Acta*, **251**: 157.

194. O'Riordan, D. M. T., Wallace, G. G. (1985), *Anal. Proc.*, **22**: 199.

195. O'Riordan, D. M. T., Wallace, G. G. (1986), *Anal. Chem.*, **58**: 128.

196. Kazuhary, S., Shumitz, T., Metsuhico, T. (1991), *J. Electroanal. Chem.*, **304**: 249.

197. Lubert, K.-H., Schnurrbusch, M. (1984), In *Euroanalysis. V*: 5th Eur. Conf. Anal. Chem. Crakow, 26–31 Aug. 1984 Book Abstr. 41.

198. Hulin, Li., Zhizing, S., Ming, Y., Jianzhong, F. (1986), *Fensi Huasue, Anal. Chem.*, **14**: 85

199. Baldwin, R. P., Kryger, L., Christensen, J. K. (1986), Abstract Paper, Pittsburgh Conference and Exposition. Anal. Chem. and Applied Spectrosc., 10–14 March 1986, Atlantic City, NJ.

200. Baldwin, R. P., Christensen, J. K., Kryger, L. (1986), *Anal. Chem.*, **58**: 1790.

201. Gehron, M. J., Brajter-Tohn, A. (1986), *Anal. Chem.*, **58**: 1488.

202. Egashira, N., Kudo, M., Hori, F. (1986), *Chem. Lett.*, **7**: 1045.

203. Kalcher, K. (1986), *Fr. Z. Anal. Chem.*, **325**: 181.

204. Kalcher, K. (1986), *Fr. Z. Anal. Chem.*, **325**: 186.

205. Lubert, K.-H., Schnurrbusch, M. (1986), *Anal. Chim. Acta*, **186**: 57.

206. Thomas, K. N., Kryger, L., Baldwin, R. P. (1988), *Anal. Chem.*, **60**: 151.

207. Jin, L., Li, Y., Fang, Y. (1988), *Fensi Huasue, Anal. Chem.*, **16**: 97.

208. Wang, Y., Dong, S. (1988), *Fensi Huasue, Anal. Chem.*, **16**: 216, 235.

209. Wang, J., Martinez, T. (1988), *Anal. Chim. Acta*, **207**: 95.

210. Martinez, T., Wang, J. (1988), Abstracts of the Pittsburgh Conference and Exposition on Analytical Chemistry and Applied Spectroscopy, 22–26 February 1988, New Orleans, LA.

211. Jin, L.-T., Shan, Y., Tong, W., Fang, Y.-Z. (1989), *Microchim. Acta*, **1**: 97.

212. Wang, J., Lin, M. S. (1988), *Anal. Chem.*, **60**: 1545.

213. Cardea-Torresdey, J., Darnell, D., Wang, J. (1988), *J. Electroanal. Chem.*, **252**: 197.

214. Dong, S., Wang, Y. (1988), *Anal. Chim. Acta*, **212**: 341.

215. Lloso Torres, J. M., Ruf, H., Schorb, K., Qche, H. J. (1988), *Anal. Chim. Acta*, **211**: 317.

216. Brainina, Kh. Z., Chernysheva, A. V., Nikitina, N. A., Stozhko, N. Yu. (1989), In Proceedings of the Third Beijing Conference and Exhibit on Instrum. Analysis, 27–30 October 1989, Beijing, China, F111–F112.

217. Wang, J., Bonakdar, M., Pack, M. M. (1987), *Anal. Chim. Acta*, **192**: 215.

218. Hayer, B., Florence, T. M. (1987), *Anal. Chem.*, **59**: 2839.

219. Wang, J., Greene, B., Morgan, C. (1984), *Anal. Chim. Acta*, **158**: 15.

220. Cox, J. A., Kulesza, P. J. (1983), *Anal. Chim. Acta*, **154**: 71.

221. Kalcher, K. (1986), *Fr. Z. Anal. Chem.*, **324**: 47.

222. Guadalupe, A. R., Jhaveri, S. S., Liu, K. E., Abruna, H. D. (1987), *Anal. Chem.*, **59**: 2436.

223. Hernandez, L., Hernandez, P., Blanco, M. H., Sanchez, M. (1988), *Analyst*, **113**: 41.

224. Kalcher, K. (1985), *Anal. Chim. Acta*, **177**: 175.

225. Kalcher, K., Greschonig, A., Pietsh, K. (1987), *Fr. Z. Anal. Chem.*, **327**: 513.

226. Dick, R., Ruf, H., Ache, H. J. (1988), Abstracts of the Pittsburgh Conference and Exposition on Analytical Chemistry and Applied Spectroscopy, 22–26 February 1988, New Orleans, LA.

227. Wang, J., Ruiliang, L. (1989), *Talanta*, **36**: 279.

228. Yoshida, H., Tanaka, S., Mitsuhiko, J. (1980), *Bunseki Kagaku*, **29**: 749.

229. Wahdaf, F., Neeb, R. (1984), *Fr. Z. Anal. Chem.*, **318**: 334.

230. Xiaohua, L., Qiantao, C., Xuolong, Y., Wenzhao, S. (1991), *J. Huazhong (Cent. China) Univ. Sci. and Technol.*, **19**: 161.

231. Wang, J., Sun, C. (1990), *J. Electroanal. Chem.*, **291**: 59.

232. Monien, H., Linke, K. (1970), *Fr. Z. Anal. Chem.*, **250**: 173.

233. Specker, H., Monien, H., Lendermann, B. (1971), *Chem. Anal. (PRL)*, **17**: 1003.

234. Brainina, Kh. Z., Tchernysheva, A. V. (1974), *Talanta*, **21**: 287.

235. Braun, H., Metzger, M. (1984), *Fr. Z. Anal. Chem.*, **318**: 321.

236. Cox, J. A., Kulesza, P. J. (1984), *Anal. Chem.*, **56**: 1021.

237. Constant, M. G., Van den Berg, R., Kevin, M., John, P. K. (1986), *Anal. Chim. Acta*, **188**: 177.

238. Jin, L., Xu, J., Fang, Y. (1986), *Fensi Huasue, Anal. Chem.*, **14**: 513.

239. Jhao, J., Jim, W. (1988), *J. Electroanal. Chem.*, **256**: 181.

240. Brainina, Kh. Z., Tchernysheva, A. V., Stozhko, N. Yu. (1984), *Zavodsk. Lab.*, **50**: 3.

241. Brainina, Kh. Z., Khanina, R. M., Stozhko, N. Yu., Tchernysheva, A. V. (1984), *Zh. Anal. Khim.*, **39**: 2068.

242. Vydra, F., Nghi, T. V. (1977), *J. Electroanal. Chem.*, **78**: 167.

243. Monien, H., Gerlach, U. (1981), *Anal. Chem.*, **306**: 136.

244. Thornton, D. C., Corby, K. T., Spendel, V. A. (1985), *Anal. Chem.*, **57**: 150.

245. Hershenhart, E., McGreery, R. L., Knight, R. D. (1984), *Anal. Chem.*, **56**: 2256.

246. Edmonds, T. R., Guoliang, Ji. (1983), *Anal. Chim. Acta*, **151**: 99.

247. Engstrom, R. C., Stresser, V. A. (1984), *Anal. Chem.*, **56**: 136.

248. Ravichanckran, K., Baldwin, R. P. (1984), *Anal. Chem.*, **56**: 1744.

249. Wang, J., Martinez, T., Janiv, D. R., McCormick, L. D. (1990), *J. Electroanal. Chem.*, **278**: 379.

250. Cross, M., Jordan, J. (1984), *Pure Appl. Chem.*, **56**: 1096.

251. Gunasingham, H., Fleet, B. (1982), *Analyst*, **107**: 896.

252. Farsang, G. (1984), In *Modern Trends Anal. Chem.*, PTA-B:A415 Budapest.

253. Bond, A. M., Fleischman, M., Khoo, S. B., Pons, S. (1986), *Indian J. Technol.*, **24**: 492.

254. Wang, J. (1982), *Anal. Chem.*, **54**: 221.

255. Gonon, F. G., Fombarlet, C. M., Buda, M. T., Pujol, T. F. (1981), *Anal. Chem.*, **53**: 1386.

256. Howell, J. O., Wightman, M. R. (1984), *Anal. Chem.*, **56**: 524.

257. Sleszynski, N., Osteryoung, J., Carter, M. (1984), *Anal. Chem.*, **56**: 130.

258. Sternitzke, K. D., McGreery, R. L. (1990), Abstract No. 307, Pittsburgh Conference and Exposition on Analytical Chemistry and Applied Spectroscopy, 5–9 March 1990, New York.

259. Sholz, F., Kupfer, M., Seelisch, J., Glowacz, G., Henrion, G. (1987), *Fr. Z. Anal. Chem.*, **326**: 774.

260. Stromberg, A. G., Kosukhin, V. A., Kuleshov, V. I. (1983), *Zavodsk. Lab.*, **49**: 24.

261. Blaedel, W. J., Wang, J. (1979), *Anal. Chem.*, **51**: 1724.

262. Bond, A. M., Hudson, H. A., Van den Bosch, P. A. (1984), *Anal. Chim. Acta*, **127**: 121.

263. Kavel, H., Umland, F. (1983), *Fr. Z. Anal. Chem.*, **316**: 386.

264. Jagner, D. (1983), *Modern Trends Anal. Chem.*, **2**: 53.

265. Jagner, D., Aren, K. (1984), *Modern Trends Anal. Chem.*, **3**: 317.

266. Ivaska, A., Ryan, T. H. (1981), *Collect Czech. Chem. Commun.*, **46**: 187.

267. Konanur, N. K., Van Loon, G. W. (1977), *Talanta*, **24**: 184.

268. Briner, R. C. et al. (1985), *Anal. Chim. Acta.*, **172**: 31.

269. Ostrovidov, E. A. (1982), *Zh. Anal. Khim.*, **37**: 1703.

270. Alzand, I. R., Langforal, C. H. (1985), *Can. J. Chem.*, **63**: 643.

271. Wang, J., Frelha Bassam, A. (1982), *Anal. Chem.*, **54**: 334.

272. Kaplin, A. A., Geineman, A. E., Saraeva, V. E., Katyukhin, V. E. (1981), *Zh. Anal. Khim.*, **36**: 1903.

273. Igolinski, V. A., Guryanova, O. N., Kotova, N. A., Zheleznyak, G. I., Noskova, G. G. (1978), In Proceedings VII, All-Union Meeting on Polarography, p. 157.

274. Stulik, K., Pacakova, V. (1981), *J. Electroanal. Chem.*, **129**: 1.

275. Wang, C. L., Brenda, R. S. (1990), Abstract Book 304, Pittsburgh Conference and Exposition on Analytical Chemistry and Applied Spectroscopy, 5–9 March 1990, New York.

276. Zarubina, R. N., Kaplin, A. A., Kolpakova, N. A. (1971), *Zavodsk. Lab.*, **37**: 5.

277. Toropova, V. F., Polyakov, Yu. I., Naumova, E. A. (1975), *Zh. Anal. Khim.*, **30**: 1187.

278. Toropova, V. F., Polyakov, Yu. I., Naumova, E. A., Kopylova, O. V. (1980), *Zh. Anal. Khim.*, **35**: 296.

279. Toropova, V. F., Polyakov, Yu. I., Zhdanova, G. N. (1983), *Zh. Anal. Khim.*, **38**: 238.

280. Bersier, P. M., Bersier, J. (1988), *Analyst*, **113**: 3.

281. Kalvoda, R. (1982), *Anal. Chim. Acta*, **138**: 11.

282. Malachova, N., Chernysheva, A., Brainina, Kh. Z. (1987), *J. Anal. Chem.*, **42**: 1636.

283. Newton, M. P., Van der Berg, C. M. (1987), *Anal. Chim. Acta*, **199**: 59.

284. Gammelgaard, B., Anderson, I. R. (1985), *Analyst*, **110**: 1197.

285. Cotzee, R. C., Albertonni, K. (1983), *Anal. Chem.*, **55**: 1516.

286. Boi, Y., Zhou, D. (1987), *Chem. J. Chin. Univ.*, **8**: 595.

287. Ganzalez, M., Gonzalez, P. C., Hernandez-Mendez, J. (1986), *Quim. Real Soc. Esp. Quim.*, **82**: 71.

288. Wang, J., Zadeil, J. M. (1986), *Anal. Chim. Acta*, **185**: 229.

289. Yao, X., Li, L., Yuan, X. (1986), *Anal. Chem. (China)*, **14**: 502.

290. Yang, X., Jackwerth, E. (1986), *Fr. Z. Anal. Chem.*, **325**: 110.

291. Wang, J., Mahmoud, J. S. (1986), *J. Electroanal. Chem.*, **208**: 383.

292. Gemmer-Colos, V., Neeb, R. (1986), *Naturwissenschalten*, **73**: 498.

293. Wang, J., Zadeil, J. (1987), *Talanta*, **34**: 909.

294. Wang, J., Tuzchi, P., Varughase, K. (1987), *Talanta*, **34**: 561.

295. Prabhu, S., Baldwin, R., Kryger, L. (1987), *Anal. Chem.*, **59**: 1074.

296. Prabhu, S., Baldwin, R., Kryger, L. (1989), *Electroanalysis*, **1**: 13.

297. Torrance, K., Catford, C. (1987), *Talanta*, 34: 939.

298. Gammelgard, B. (1986), *Anal. Proc.*, 23: 222.

299. Magyor, Bl., Wundoreli, H. (1985), *Microchim. Acta*, 3: 223.

300. Sun, C., Wang, J., Hu, W., Mao, X. (1991), *J. Electroanal. Chem.*, 306: 251.

301. Popkova, G. N., Fyodorova, N. D., Brainina, Kh. Z., Fokina, L. S. (1984), *Zh. Anal. Khim.*, 39: 2147.

302. Popkova, G. N., Fyodorova, N. D., Brainina, Kh. Z. (1985), *Zavodsk. Lab.*, 51: 3.

303. Bläzquez, L. C., Garcia-Monco, R. M., Cabanillas, A. G., Sanchez, M. A. (1989), *Fr. Z. Anal. Chem.*, 334: 166.

304. Zhao, Z., Cai, X., Li, P. (1987), *Talanta*, 34: 813.

305. Zonggui, X., Sanhong, F. (1987), *Fensi Huasue, Anal. Chem.*, 15: 926.

306. Chen, Q., Wei, R. (1987), *Tunzsi Dasue Suebao, J. Tongji Univ.*, 15: 355.

307. Gao, J., Xu, M., Pan, Z. (1987), *Uhang Gasue Suebao, J. Wuhon Univ. Natur., Sci. Ed.*, 2: 75.

308. Wang, J., Mahmoud, J. (1987), *Fr. Z. Anal. Chem.*, 327: 789.

309. Wahdat, F., Neeb, R. (1985), *Fr. Z. Anal. Chem.*, 320: 334.

310. Kavel, H., Umland, F. (1986), *Fr. Z. Anal. Chem.*, 325: 191.

311. Donal, J. R., Bruland, K. W. (1988), *Anal. Chem.*, 60: 240.

312. Wang, J., Varughese, K. (1987), *Anal. Chim. Acta*, 199: 185.

313. Jering, H. (1974), *Electrosorptionanalysise mit der Wechsel-strompolarographie*, Akademie-Verlag, Berlin.

314. Kazarinov, V. E., Khorani, D., Vasiliev, Yu. B., Andreev, V. N. (1985), *Elektrokhim.*, 22: 97.

315. Frumkin, A. N. (1979), *Zero Charge Potentials*, Nauka, Moscow.

316. Krznaric, D., Valenta, P., Nürenberg, H. W. (1975), *J. Electroanal. Chem.*, 65: 863.

317. Valenta, P., Nürnberg, H. W., Krznaric, D. (1976), *Bioelectrochem. Bioenerg.*, 3: 418.

318. Krznaric, D., Valenta, P., Nürnberg, H. W., Branica, M. (1978), *J. Electroanal., Chem.*, 93: 41.

319. Temerk, Y. M., Valenta, P., Nürnberg, H. W. (1979), *J. Electroanal. Chem.*, 100: 77.

320. Palecek, E. (1980), in Smyth, W. F. (Ed.), *Electroanalysis in Hygiene, Environmental, Clinical and Pharmaceutical Chemistry*, Elsevier, Amsterdam.

321. Nürnberg, H. W., Valenta, P. (1976), *Croat. Chem. Acta*, 48: 623.

322. Valenta, P., Nürnberg, H. W., Klahre, P. (1975), *Bioelectrochem.Bioenerg.*, 2: 204.

323. Sequaris, J. M., Valenta, P., Nürnberg, H. W., Malfoy, B. (1978), *Bioelectrochem. Bioenerg.*, 5: 483.

324. Palecek, E., Jelen, F., Manousek, O. (1980), *Collect. Czech. Chem. Commun.*, 45: 3460.

325. Homolka, J., Proceedings J. Heyrovsky Memorial Congress on Polarography, Prague, Czechoslovakia, Aug. 25–29, 1980, Vol. 1.

326. Volke, J., Proceedings J. Heyrovsky Memorial Congress on Polarography, Prague, Czechoslovakia, Aug. 25–29, 1980, Vol. 1.

327. Jelen, F., Paleček, E., Manousek, O. (1980), Third Bruno Symposium on Molecular Biophysics, Czechoslovakia, p. 43.

328. Paleček, E., Kolar, V., Jelen, F., Heineman, U. (1990), *Bioelectrochem. Bioenerg.*, **23**(3): 285.

329. Paleček, E., Jelen, F. (1980), *Collect. Czech. Chem. Commun.*, **45**: 3472.

330. Paleček, E. (1980), *Anal. Biochem.*, **108**: 129.

331. Paleček, E. (1980), *Anal. Lett.*, **13**(B5): 331.

332. Brabec, V. (1981), *Bioelectrochem. Bioenerg.*, **8**: 437.

333. Kuznetsov, V. A., Shumakovich, G. P. (1975), *Methods of Modern Biochemistry*, (in Russian), Nauka, Moscow, p. 160.

334. Kuznetsov, V. A., Shumakovich, G. P. (1989), in Bard, A. J. (Ed.), *Electroanalytical Chemistry*, Vol. 16, p. 105, Marcel Dekker, New York.

335. Brown, A. P., Koval, C., Anson, F. (1976), *J. Electroanal. Chem.*, **72**: 379.

336. Brainina, Kh. Z., Chernysheva, A. V. (1977), *Polarography: Problems and Prospects Zinatne*, Riga, p. 242.

337. Househam, B. C., Vanden Berg, C. M., Riley, J. P. (1987), *Anal. Chim. Acta*, **200**: 291.

338. Baldwin, R., Packett, R., Woodcock, T. M. (1981), *Anal. Chem.*, **53**: 540.

339. Zoulis, N. E., Efstathiou, C. E. (1988), *Anal. Chim. Acta*, **204**: 207.

340. Cataldi, T. R., Guerrieri, A., Palmisano, F., Lambonin, P. G. (1988), *Analyst*, **113**: 869.

341. Barec, J., Balsiene, J., Berka, A., Hauserova, Z. (1988), *Collect. Czech. Chem. Commun.*, **53**: 19.

342. Hernandez, L., Zapradeil, A., Antonio, J., López, P., Bermejo, F. (1987), *Analyst*, **112**: 1149.

343. Tocksteinova, Z., Kopanika, M. (1987), *Anal. Chim. Acta*, **199**: 77.

344. Wang, J., Taha, Z. (1988), *Anal. Chim. Acta*, **215**: 29.

345. Lorenzo, E., Hernandez, L. (1987), *Anal. Chim. Acta*, **201**: 275.

346. Kopanica, M., Stara, V. (1986), *J. Electroanal. Chem.*, **214**: 115.

347. Sun, S., Hu, W., Shao, J., Zhu, Y. (1986), *Chem. J. Chin. Univ.*, **7**: 786.

348. Wang, J., Lin, M. S., Villa, V. (1987), *Analyst*, **112**: 1303.

349. Kitamura, H., Yamada, Y., Nakamoto, M. (1984), *Chem. Lett.*, **6**: 837.

350. Wang, J., Ferrias, P. A., Machmoud, J. S. (1986), *Analyst*, **111**: 837.

351. Barek, J., Pastor, T. J., Volavova, S., Zima, J. (1987), *Collect. Czech. Chem. Commun.*, **52**: 2149.

352. Siria, J. W., Baldwin, R. P. (1980), *Anal. Lett. A Chem. Anal.*, **13**: 577.

353. Ciszkowska, M., Stojek, Z. (1984), In Euroanalysis V: 5th Eur. Conf. Anal. Chem. Cracow, Aug. 26–31, 1984, Abstr. Book Cracow.

354. Wang, J., Mahmoud, J. S. (1986), *Anal. Chim. Acta*, **186**: 31.

355. Fernandez, A. J. M., Costa Garsia, A., Miranda, O. A. J., Tunon, B. P. (1987), *J. Electroanal. Chem.*, **225**: 241.

356. Hernandez, L., Hernandez, P., Blanco, M. H., Lorenzo, E., Alda, E. (1988), *Analyst*, **113**: 1719.

357. Ordieres, A. J. M., Gutieris, M. J. G., Garsia, A. C., Blanko, P. T. (1987), *Analyst*, **112**: 243.

358. Tanaka, S., Yoshida, H. (1988), *Talanta*, **35**: 837.

359. Wang, J., Tuzhi, P., Lin, M.-S., Tapia, T. (1986), *Talanta*, **33**: 707.

360. Temizer, A., Nur Onar, A. (1988), *Talanta*, **35**: 805.

361. Wang, J., Lin, M. S., Villa, V. (1986), *Anal. Lett.*, **19**: 2293.

362. O'Dea, Ph., Garcia, A. C., Ordieres, A. J. M., Blanko, P., Smyth, M. R. (1991), *Electroanalysis*, **3**: 337

363. Pinilla, G. F., Calvo Blazquez, L., Garcia-Monco Corra, R. M., Sanchez Misiego, A. (1988), *Fr. Z. Anal. Chem.*, **332**: 821.

364. Sivakumar, A., Jayarama, R. S., Krishnan, V. R. (1988), *Trans. SAEST*, **23**: 361.

365. Diez-Caballero, R. B., Arranz, V. J. F., Carćes Moyo, J. J., Coicoba, A. A. (1988), *Analyst*, **113**: 1047.

366. Barek, J., Balsiene, J., Tictzovv, B., Zima, J. (1988), *Collect. Czech. Chem. Commun.*, **53**: 921.

367. Procopio, J. R., Escribano, M. T. S., Hernandez, L. H. (1988), *Fr. Z. Anal. Chem.*, **331**: 27.

368. Vire, J. C., Lopez, V., Patriarche, G. J., Christian, G. D. (1988), *Anal. Lett.*, **21**: 2217.

369. Kuzmin, N. M., Zolotov, Yu. A., Karbainov, Yu. A. (1969), *Proc. Comm. Anal. Chem. AN SSSR*, **17**: 288.

370. Nghi, T. V., Vydra, F. (1975), *Anal. Chim. Acta*, **80**: 267.

371. Vydra, F. (1976), *Chem. Listy*, **R70**: 327.

372. Hulanyski, A. (1985), In *Euroanalysis V. Reviews of Analytical Chemistry Budapest*, Kiado, p. 121.

373. Lexa, J., Stulik, K. (1985), *Chem. Listy*. **79**(1): 58.

374. Kaplin, A. A., Stromberg, A. G., Pikula, N. P. (1977), *Zavodsk. Lab.*, **43**: 385.

375. Brainina, Kh. Z., Nikulina, I. N., Kurbatova, V. I. (1973), *Zovodsk. Lab.*, **39**: 777.

376. Petrova, L. G., Ignatov, V. I., Neiman, E. Ya. (1978), *Zh. Anal. Khim.*, **33**: 1322.

377. Mizuike, A., Miwa, T., Fuiju, Y. (1974), *Microchim. Acta*, **4**: 595.

378. Benniaminova, S. M. (1981), *Methods of Extraction and Determination of Noble Elements*, Nauka, Moscow.

379. Temmerman, E., Verbeek, F. (1972), *Anal. Chim. Acta*, **58**: 263.

380. Gornostaeva, T. D., Pronin, V. A. (1981), *Zh. Anal. Khim.*, **36**: 295.

381. Mizuike, A., Niwa, T., Fuiju, Y. (1975), *Microchim. Acta*, **1**: 125.

382. Zebreva, A. I., Matakova, R. N., Zholdyvaeva, R. B. (1981), *Zh. Anal. Khim.*, **36**: 405.

383. Kaplin, A. A., Pichugina, V. I. (1984), *Zh. Anal. Khim.*, **39**: 664.

384. Kaplin, A. A. (1978), *Zh. Anal. Khim.*, **33**: 1972.

385. Kaplin, A. A., Zaichko, L. F., Dubova, N. M. (1977), *Zavodsk. Lab.*, **43**: 939.

386. Kaplin, A. A., Portnyagina, E. O., Gridaev, V. F. (1979), *Zh. Anal. Khim.*, **34**: 950.

387. Kaplin, A. A., Mikhailova, Z. S., Zaichenko, L. F. (1978), *Zh. Anal. Khim.*, **33**: 120.

388. Kaplin, A. A., Mamontova, I. P. (1977), *Zh. Anal. Khim.*, **33**: 703.

389. Nürnberg, H. W. (1984), *Anal. Chim. Acta*, **164**: 1.

390. Brainina, Kh. Z., Roitman, L. I., Khanina, R. M., Gruskova, N. A. (1985), *Khim. Tekhn. Vody*, **7**: 27.

391. Florence, T. M. (1986), *Analyst*, **111**: 489.

392. Brainina, Kh. Z., Khanina, R. M., Roitman, L. I. (1985), *Anal. Lett.*, **18**(A2): 117.

393. Salberg, P., Lund, W. (1982), *Talanta*, **29**: 457.

394. Zakharova, E. A., Mokrousov, G. M., Volkova, V. N., Lisetski, V. N. (1983), *Zh. Anal. Khim.*, **38**: 1584.

395. Clem, R. C., Radson, A. J. (1978), *Anal. Chim.*, **50**: 102.

396. Florence, T. M. (1982), *Talanta*, **29**: 345.

397. Zirino, A., Lieberman, S. H., Clavell, C. (1978), *Environ. Sci. Technol.*, **12**: 73.

398. Jaya, S., Rao, P., Rao, P. G. (1985), *Talanta*, **32**: 1061.

399. Eskilssen, H., Jagner, D. (1982), *Anal. Chim. Acta*, **138**: 27.

400. Mart, L., Nürnberg, H. W., Dyrssen, D. (1981), in *Trace Metals Sea Water*, Proc. NATO Adv. Res. Inst. Erice, 30 March–3 April 1981, New York.

401. Lund, W., Salberg, M. (1975), *Anal. Chim. Acta*, **76**: 131.

402. Miwa, T., Mizuike, A. (1977), *Bunseki Kagaku* **26**: 588.

403. Obiols, T., Libere, L. (1984), *Affinidad*, **41**: 315.

404. Arts, W., Bleitschneider, H., Rickert, B. (1984), *Fr. Z. Anal. Chem.*, **319**: 501.

405. Sadana, R. S. (1983), *Anal. Chem.*, **55**: 304.

406. Xuewen, G. (1983), *Fensi Kuekeju*, **5**: 362.

407. Pea, V. (1977), *Wasser*, **48**: 89.

408. Bonelli, J. E., Taylor, H. E., Scogerbve, R. K. (1980), *Anal. Chim. Acta*, **118**: 243.

409. Clark, B. R., Depaoli, D. W., McTaggart, D. R., Palton, B. D. (1988), *Anal. Chim. Acta*, **215**: 13.

410. Nagourney, S. I., Bogen, D. C. (1978), *Can. I. Spectrosc.*, **23**: 101.

411. Nürnberg, H. W. (1985), *Anal. Proc.*, **22**: 378.

412. Nangniot, P. (1985), *Trends Anal. Chem.*, **4**: 155.

413. Kalvoda, R., Novony, L. (1986), *Collect. Czech. Chem. Commun.*, **51**: 1581.

414. Morgil, F. I. (1986), *Chromatogr methody a vyznanu zdravie cloveka*, Zb suhrnov 7 Konf Talranska Lomnica, 18–20 Nov. 1984.

415. Alonso, R. M., Jimenez, R. M., Fogg, A. G. (1988), *Analyst*, **133**: 27.

416. Sestákove, I., Kopanica, M. (1988), *Talanta*, **35**: 846.

417. Gawargioris, V. A., Tadros, N. B., Besada, A., Ibrahim, L. F. (1987), *Analyst*, **112**: 549.

418. Chastel, O., Kaufmann, J.-M., Patriarche, G. J., Christian, G. D. (1989), *Anal. Chem.*, **61**: 170.

419. Ivaska, A., Vanevsorn, Y., Davidson, I. E., Smyth, W. F. (1980), *Anal. Chim. Acta*, **121**: 51.

420. Christenssen, I. K., Kryger, L., Pind, N. (1982), *Anal. Chim. Acta*, **141**: 131.

421. Cammann, K. (1977), *Int. Symp. Microchim. Tech.*, *Abstr. Davos*, SI.

422. Locatelli, C. (1984), *Magy Kém folioirat*, **90**: 448.

423. Slepushkin, V. V., Kuzmina, N. N., Yartsev, M. G. (1975), *Izv. Vuzov. Khim. i Khim. Teckhnol.*, **18**: 1391.

424. Weber, G. (1986), *Fr. Z. Anal. Chem.*, **322**: 311.

425. Batley, G. E., Farrar, Y. (1978), *Anal. Chim. Acta*, **99**: 292.

426. Matson, W. R. (1977), Abstracts of the Pittsburgh Conference and Exposition on Analytical Chemistry and Applied Spectroscopy, Cleveland, OH.

427. Brzezinska, A., Trzozinska, A. (1983), *Chem. Anal. (PRL)*, **28**: 121.

428. Rongshon, Lü., Zhoubo, S. (1988), *Fensi Huasue, Anal. Chem.*, **16**: 179.

429. Fogg, A. G., Fleming, R. M. (1987), *Port. Electrochim. Acta*, **5**: 299.

430. Kalvoda, R., Kopanica, M. (1989), *Pure and Appl. Chem.*, **61**: 97.

431. Neiman, E. Ya., Dracheva, L. V. (1990), *J. Anal. Chem. (USSR)*, **45**: 222.

432. Borrachero, A., Rodriguez, J., Vinagre, F., Sánchez, A. (1989), *Analyst*, **113**: 1795.

433. Rodriguez, J., Vingre, F., Borrachero, A., Sánchez, A. (1989), *Analyst*, **114**: 393.

434. Rodriguez, J., O'Kennedy, R., Smyth, M. R. (1988), *Anal. Chim. Acta*, **212**: 355.

435. Brainina, Kh. Z., Khodos, M. Ya., Belisheva, G. M., Vidrevich, M. B. (1990), *Z. fur Phys. Chem. Nue Folge.*, **168**: 65.

436. Bauer, D., Gaillochet, M. P. (1974), *Electrochem. Acta*, **19**: 597.

437. Vidrevich, M. B., Uritskaya, A. A., Kitaev, G. A. (1984), *Zavodsk Lab.*, **50**: 17.

438. Zakharchuk, N. F., Valisheva, N. A., Yudelevich, N. G. (1980), *Izv. SO AN SSSR, Ser. Khim. Nauk.*, **4/2**: 39.

439. Clicksmann, R., Morehause, C. K. (1958), *J. Electrochem. Soc.*, **105**: 299.

440. Clicksmann, R., Morehause, C. K. (1958), *J. Electrochem. Soc.*, **105**: 613.

441. Clicksmann, R., Morehause, C. K. (1959), *J. Electrochem. Soc.*, **106**: 741.

442. Sapozhnikova, E. Ya., Roizenblat, E. M., Safonova, S. V., Bezrukov, V. I., Karpov, V. P. (1976), *Elektrokhim.*, **12**: 1730.

443. Rozhdestvenskaya, Z. B., Muldagalieva, I. Kh., Zharmenov, A. A. (1982), *Electrokhim.*, **18**: 122.

444. Muldagalieva, I. Kh., Rozhdestvenskaya, Z. B., Songina, O. A. (1973), *Elektrokhim.*, **9**: 796.

445. Krapivkina, T. A., Noacheva, V. V. (1977), *Zavodsk. Lab.*, **43**: 263.

446. Shibalko, G. V., Brainina, Kh. Z., Stenina, N. I. (1981), *Zavodsk. Lab.*, **47**: 10.
447. Zakharchuk, N. F., Mustafina, G. F., Smirnova, T. P., Illarionova, I. S. (1988), *Elektrokhim.*, **24**: 26.
448. Medvedeva, E. N., Rozhdestvenskaya, Z. B., Kulikovski, A. K. (1976), *Izv. AN Kaz SSR, Ser. Khim.*, **4**: 80.
449. Marschakov, I. K., Tutukina, N. M., Stepushkin, N. I. (1977), *Zavodsk. Lab.*, **43**: 974.
450. Nysanbaeva, Z. A., Nyerbekov, B. Yu., Bedembaev, G. E., Khisametdinov, A. M. (1988), *Izv. AN Kaz SSR, Ser. Khim.*, **3**: 32.
451. Pastukhov, V. P., Frishberg, I. V., Kalnishevskaya, L. N., Brainina, Kh. Z., Roitman, L. I. (1983), *Poroshk Metallurgiya*, **5**: 67.
452. Matakova, R. N., Yulmetova, R. F., Zebreva, A. I. (1984), *Zh. Anal. Khim.*, **39**: 1200.
453. Xiang, M. A. (1988), *J. Electroanal. Chem.*, **242**: 63.
454. Xiang, M. A., Luo, S., Zhao, H. (1988), *Chin. J. Met. Sci. Technol.*, **4**: 228.
455. Scholz, F., Nitschke, L., Henrion, G. (1990), *Electroanalysis*, **2**: 85.
456. Brage, M., Lamache, M., Bauer, D. (1979), *Electrochim. Acta*, **24**: 25.
457. Brainina, Kh. Z., Lesunova, R. P., Serebryakova, L. N. (1974), *Zavodsk. Lab.*, **49**: 632.
458. Vidrevich, M. B., Uritskaya, A. A., Khaimova, I. M., Kitaev, G. A. (1983), *Izv. AN SSSR Ser. Neorgan. Mater.*, **19**: 16.
459. Kitaev, G. A., Uriskaya, A. A., Vidrevich, M. B. (1984), *Izv. AN SSSR Ser Neorgan. Mater.*, **20**: 17.
460. Vidrevich, M. B., Drozdova, T. A., Kitaev, G. A. (1983), *Zavodsk. Lab.*, **49**: 7.
461. Vidrevich, M. B., Slinkina, M. Y., Kitaev, G. A. (1983), *Izv. AN SSSR Ser. Neorgan. Mater.*, **19**: 1257.
462. Vidrevich, M. B., Uritskaya, A. A., Drozdova, T. A., Kitaev, G. A. (1984), *Izv. AN SSSR Ser. Neorgan. Mater.*, **20**: 341.
463. Sapozhnikova, E. Ya., Roizenblat, E. M., Safonova, S. V. (1975), *Zavodsk. Lab.*, **41**: 1189.
464. Lunev, M. I., Kamenev, A. I., Agasyan, P. K. (1975), *Vestn. Mosk. Un-ta, Khim.*, **16**: 586.
465. Kamenev, A. I., Lunev, M. I., Agasyan, P. K. (1977), *Zh. Anal. Khim.*, **32**: 550.
466. Muldagalieva, I. K., Rozhdestvenskaya, Z. B., Oskanova, F. K. (1979), In *Study of Heterogeneous Systems*, Kaz Gu Publishing House, Alma-Ata.
467. Songina, O. A. (1978), *Talanta*, **25**: 116.
468. Rozhdestvenskaya, Z. B., Sigatov, V. V. (1979), *Elektrokhim.*, **15**: 1530.
469. Gruner, W., Stahlberg, R., Brainina, Kh., Akselrod, N., Kamyshov, V. (1990), *Electroanalysis*, **2**: 397.
470. Songina, O. A., Rozhdestvenskaya, Z. B., Smirnova, V. V. (1980), *Izv. Vuzov. Khim. i Khim. Tekhnol.*, **23**: 1507.

471. Zakharchuk, N. F., Valisheva, N. A., Yudelevich, I. G. (1980), *Izv. SO AN SSSR, Ser. Khim.*, **4/2**: 45.

472. Bely, V. I., Zakharchuk, N. F., Smirnova, T. P., Yudelevich, I. G. (1980), *Elektronnaya Promyshlennost*, **11–12**: 35.

473. Smirnova, T. P. (1981), *Thin Solid Films*, **76**: 11.

474. Smirnova, T. P., Shpurik, V. N., Bely, V. I., Zakharchuk, N. F. (1982), *Izv. SO AN SSSR, Ser. Khim.*, **6**: 93.

475. Zakharchuk, N. F., Gruner, W., Illarionova, I. S., Potapova, O. G., Lelkin, K. P. (1988), *Elektrokhim.*, **24**: 667.

476. Songina, O. A. (1973), *Elektrokhim.*, **9**: 1310.

477. Vidrevich, M. B., Petrov, A. N., Zhukovski, V. M. (1980), *Izv. AN SSSR, Ser. Neorgan. Mater.*, **16**: 1789.

478. Grigorieva, M. F., Kots, E. A. (1989), *Problems of Modern Analytical Chemistry*, (*Leningrad*), **6**: 97.

479. Moskvin, L. N., Kots, E. A., Grigorieva, M. F. (1988), *Zh. Anal. Khim.*, **43**: 1025.

480. Watson, N. V., Scott, A. V. (1965), *J. Electrochem. Soc.*, **112**: 883.

481. Ide, M., Kishi, T., Nagai, T. (1973), *Electrochem. Soc. Jpn.*, **41**: 650.

482. Miura, T., Kishi, T., Nagai, T. (1978), *Electrochem. Soc. Jpn.*, **46**: 402.

483. Mokerov, V. G., Sigalov, B. L. (1972), *Fiz. Tverd. Tela.*, **14**: 3405.

484. Shmukler, Yu. S., Kuzmin, L. L. (1965), *Protective Metallic and Oxide Coatings, Corrosion of Metals, and Studies in the Field of Electrochemistry*, Nauka, Moscow.

485. Dzharkesheva, Z. T., Zakharov, V. A., Bekturova, G. B., Songina, O. A. (1978), *Izv. AN Kaz SSR, Ser. Khim.*, **3**: 30.

486. Grigorieva, M. F., Zakharchenko, V. M., Tserkovnitskaya, I. A., Denshova, I. A. (1980), *Zavodsk. Lab.*, **46**: 24.

487. Bazarova, E. V., Khodos, M. Ya., Brainina, Kh. Z. (1984), *Elektrokhim.*, **20**: 609.

488. Pletnev, R. N., Ivakin, A. A., Gorshkov, V. V., Chirkov, A. K. (1975), *Dokl. AN SSSR*, **224**: 106.

489. Fiermans, L., Clauws, P., Lambrecht, W. (1980), *Phys. Stat. Solidi (a)*, **59**: 485.

490. Brainina, Kh. Z., Bazarova, E. V., Khodos, M. Ya. (1984), *Elektrokhim.*, **20**: 613.

491. Vetter, K. J. (1961), *Elektrochemische Kinetik*, Springer, Berlin.

492. Vetter, K. J., Jager, N. (1966), *Electrochim. Acta*, **11**: 401.

493. Teltow, I. (1949), *Ann. Phys.*, **5**: 63.

494. Tarasevich, M. R., Khruschova, E. I., Shumilova, N. A. (1978), *Final Results of Science and Engineering*, Vol. 13, Elektrokhim, VINITI, Moscow.

495. Brainina, Kh. Z., Bazarova, E. V., Volkov, V. L. (1980), *Elektrokhim.*, **16**: 1203.

496. Bazarova, E. V., Volkov, V. L., Brainina, Kh. Z. (1980), *Elektrokhim.*, **16**: 825.

497. Trassatti, S., Lodi, G. (1981), in Trassatti, S. (Ed.), *Electrodes of Conductive Metallic Oxides*, Part B, p. 521, Elsevier, Amsterdam.

498. Sapozhnikova, E. Ya., Davidovich, A. G., Roizenblat, E. M. (1982), *Zh. Neorgan. Khim.*, **27**: 2888.

499. Ruby, W. R., Tremmel, C. G. (1968), *J. Electroanal. Chem.*, **18**: 231.

500. Sapozhnikova, E. V. (1981), *Zh. Neorgan. Khim.*, **26**: 1751.

501. Lecuire, J.-M., Evrard, O. (1977), *J. Electroanal. Chem.*, **78**: 331.

502. Lecuire, J.-M. (1975), *J. Electroanal. Chem.*, **66**: 195.

503. Sapozhnikova, E. Ya., Roizenblat, E. M., Klimov, V. V. (1974), *Izv. AN SSSR, Ser. Neorgan. Mater.*, **10**: 1085.

504. Sapozhnikova, E. Ya., Roizenblat, E. M., Titenko, A. G., Gurevich, L. V. (1976), *Izv. AN SSSR Ser. Neorgan. Mater.*, **12**: 2236.

505. Brainina, Kh. Z., Khodos, M. Ya., Vidrevich, M. B. (1985), *Voltammetry of Organic and Inorganic Compounds*, p. 221, Nauka, Moscow.

506. Kiselev, V. D., Krylov, O. V. (1979), *Electronic Phenomena in Adsorption and Catalysis on Semiconductors*, Nauka, Moscow.

507. Sapozhnikova, E. Ya., Roizenblat, E. M., Klimov, V. V., Maslova, V. I. (1978), *Elektrokhim.*, **14**: 1630.

508. Hodos, M. Ya., Bazarova, E. V., Palkin, A. P., Brainina, Kh. Z. (1984), *J. Electroanal. Chem.*, **164**: 121.

509. Brainina, Kh. Z., Bazarova, E. V., Volkov, V. L. (1980), *Elektrokhim.*, **16**: 69.

510. Vidrevich, M. B., Bamburov, V. G., Zhukovski, V. M. (1979), *Izv. AN SSSR, Ser. Neorgan. Mater.*, **15**: 2251.

511. Vidrevich, M. B., Zhukovski, V. M., Bamburov, V. G. (1979), *Izv. AN SSSR Ser. Neorgan. Mater.*, **15**: 122.

512. Tarasevich, M. R., Efremov, B. N. (1980), in Trassatti, S. (Ed.) *Electrodes of Conductive Metallic Oxides. Part A*, p. 221, Elsevier, Amsterdam.

513. Kobussen, A., Mesters, C. (1980), *J. Electroanal. Chem.*, **115**: 131.

514. Gottesfeld, S., Srinivassen, S. (1978), *J. Electroanal. Chem.*, **86**: 89.

515. Shalaginov, V. V., Belova, I. L., Roginskaya, Yu. Sh. (1979), *Elektrokhim.*, **14**: 1708.

516. Brainina, Kh. Z., Khodos, M. Ya., Belysheva, G. M., Krivosheev, N. V. (1984), *Elektrokhim.*, **20**: 1380.

517. Shinagava, M. Ea. (1968), *Rev. Polarogr.*, **15**: 108.

518. Matsumoto, Y., Kurimoto, J., Sato, E. (1979), *J. Electroanal. Chem.*, **102**: 77.

519. Pakhomov, V. I., Shub, D. M., Veselovski, V. I. (1969), *Elektrokhim.*, **5**: 843.

520. Shub, D. M., Remney, A. A., Veselovski, V. I. (1973), *Elektrokhim.*, **9**: 680.

521. Sapozhnikova, E. Ya., Roizenblat, E. M., Klimov, V. A. (1974), *Elektrokhim.*, **10**: 1284.

522. Prisedski, V. V., Sapozhnikova, E. Ya., Roizenblat, E. M. (1977), *Elektrokhim.*, **14**: 638.

523. Savin, V. S., Travina, G. Ya., Fedotov, N. A. (1970), *Elektrokhim.*, **6**: 420.

524. Vasilieva, I. A., Sukhuschina, I. S. (1980), *Vest. MGU, Ser. 2, Khimia.*, **21**: 219.

525. Hodos, M. Ya., Belisheva, G. M., Brainina, Kh. Z. (1986), *Elektrokhim.*, **22**: 493.

526. Leonidov, I. A., Fotiev, A. A., Serkalo, A. A., Hodos, M. Ya. (1987), *Zh. Neorgan. Khim.*, **32**: 1784.

527. Leonidov, I. A., Fotiev, A. A., Hodos, M. Ya. (1987), *Izv. AN SSSR, Ser. Neorgan. Mater.*, **23**: 121.

528. Sapozhnikova, E. Ya., Roisenblat, E. M., Klimov, V. V., Maslova, V. M., Kudrenko, I. A., Salei, V. S. (1978), *Elektrokhim.*, **14**: 1533.

529. Fotiev, A. A., Shulgin, B. V., Moskvin, A. S., Garvilov, F. F. (1976), *Vanadium Crystallophosphoruses*, Nauka, Moscow.

530. Hodos, M. Ya., Belisheva, G. M., Brainina, Kh. Z. (1986), *Elektrokhim.*, **22**: 488.

531. Hodos, M. Ya., Belisheva, G. M. (1985), *Elektrochemische Analysenmethoden*, Karl-Marx-Universität, Leipzig.

532. Boreskov, G. K. (1964), *Adv. Catal.*, **15**: 285.

533. Winter, E. R., Sferge, F. (1968), *J. Chem. Soc.* (A), **12**: 2889.

534. Hodos, M. Ya., Juravlev, V. D., Fotiev, A. A. (1978), *Izv. AN SSSR, Ser. Neorgan. Mater.*, **14**: 1138.

535. Fotiev, A. A., Slobodin, B. V., Khodos, M. Ya. (1988), *Vanadates: Composition, Synthesis, Structure, Properties*, Nauka, Moscow.

536. Fotiev, A. A., Gubanov, V. A. (1978), *Zh. Neorgan. Khim.*, **23**: 1898.

537. Lazukova, N. J., Gubanov, V. A., Mokerov, B. G. (1977), *Int. J. Quantum. Chem.*, **12**: 915.

538. Waltersson, K. (1976), *Chem. Commun., University of Stockholm*, **7**: 1.

539. Mokerov, V. G., Sigalov, B. D. (1972), *Fiz. Tverd. Tela.*, **14**: 3405.

540. Kofstad, P. (1972), *Nonstoichiometry, Diffusion and Electrical Conductivity in Binary Metal Oxides*, Wiley-Interscience, New York.

541. Hodos, M. Ya., Belisheva, G. M., Krivosheyev, N. V. (1988), *Zh. Neorgan. Khim.*, **33**: 1066.

542. Shalaginov, V. V., Belova, I. L., Roginskaya, Yu. Sh. (1978), *Elektrokhim.*, **14**: 1708.

543. Akselrod, N. L., Skovorodina, M. G., Vidrevich, M. B., Tkachenko, E. V., Hodos, M. Ya. (1986), *Izv. AN SSSR, Ser. Neorgan. Mater.*, **22**: 1161.

544. Krashenko, T. I., Sirneva, O. N., Fotiev, A. A. (1984), *Izv. AN SSSR, Ser. Neorgan Mater.*, **20**: 483.

545. Wadsley, A. D. (1970), *Perspectiv. Struct. Chem.*, **3**: 1.

546. Andreyeva, N. A., Gropyianov, V. M., Kozlovsky, I. V. (1969), *Izv. AN SSSR, Ser. Neorgan. Mater.*, **5**: 1302.

INDEX